HEAT

kilojoules	kilocalories	British thermal units $(10^{-5}$ therms$)$
1	0.239	0.948
4.184	1	3.968
1.054	0.252	1

TEMPERATURE

Expressed as °C (= degrees Centigrade or degrees Celsius)
Degrees Fahrenheit (°F) = 1.8(°C) + 32
$\qquad\qquad\qquad$ (°C) = 0.556(°F − 32)
Degrees Kelvin $\quad$ (°K) = (°C) + 273.15
$\qquad\qquad\qquad$ 0°K = −273.15°C = −459.7°F

INDUSTRIAL ORGANIC CHEMICALS IN PERSPECTIVE

PART ONE: RAW MATERIALS AND MANUFACTURE

OTHER BOOKS BY THE AUTHORS

The Phosphatides, by Harold A. Wittcoff, Reinhold, New York, 1950

The Chemical Economy, by Bryan G. Reuben and Michael L. Burstall, Longman, London, 1973

INDUSTRIAL ORGANIC CHEMICALS IN PERSPECTIVE

PART ONE: RAW MATERIALS AND MANUFACTURE

HAROLD A. WITTCOFF, Ph.D.

Director of R & D, Koor Chemicals Ltd., Beer-Sheva, Israel
Adjunct Professor, Department of Chemistry
University of Minnesota, Minneapolis
Vice President of Corporate Research
General Mills, Inc. (Ret.)

BRYAN G. REUBEN, M.A., D.Phil.

Principal Lecturer, Department of Chemical Engineering
Polytechnic of the South Bank, London

A WILEY-INTERSCIENCE PUBLICATION

JOHN WILEY & SONS, New York • Chichester • Brisbane • Toronto

Library of Congress Cataloging in Publication Data:

Wittcoff, Harold.
  Industrial organic chemicals in perspective.
  "A Wiley-Interscience publication."
  Includes index.
  CONTENTS: pt. 1. Raw materials and manufacture.
  1. Chemistry, Organic. 2. Chemicals—
Manufacture and industry. I. Reuben, B. G.,
joint author. II. Title.
TP247.W59   661'.8   79-19581
ISBN 0-471-03811-3

Printed in the United States of America

10 9 8 7 6 5 4 3

TO
DOROTHY AND CATHERINE

FOREWORD

The American Chemical Society study, *Chemistry in the Economy* (Washington, DC, 1973), was carried out by a committee that it was my privilege to chair jointly with my good friend, Dr. Max Tischler. In the course of the study we became acutely aware that the science and technology of chemistry directly and indirectly interact with every phase of everyday living—the food we eat, the clothes we wear, our health, welfare, housing, transportation, and national security. In addition, the study indicated that about 70% of all chemists and chemical engineers in the United States work in industry. It is surprising therefore to find the science and technology of chemistry in industry to be largely unexplored in modern curricula in our colleges and universities.

Based on that study, the American Chemical Society has recommended that more industrial courses be included in both the undergraduate and graduate chemical curricula. We are very pleased, accordingly, to have two volumes that bring the real world of industrial organic chemistry and technology into the classroom. *Industrial Organic Chemicals in Perspective* is based on courses that Dr. Wittcoff organized at the University of Minnesota and Dr. Reuben has been presenting at the University of Surrey and the Polytechnic of the South Bank in London, England. Both of the authors have extensive industrial background and experience.

It is especially fascinating to observe how these volumes trace the flow of organic chemicals through industry. For example, the petroleum refinery provides only a few basic starting materials. From these, through complex and sophisticated chemistry, much of it based on homogeneous and heterogeneous catalysis, there result billions of pounds of numerous products that give rise to a vast and important technology to which a large portion of this book is devoted.

This textbook should provide the basis for much needed courses in industrial organic chemistry in colleges and universities. Such courses will endow students entering the chemical and allied industries with a level of knowledge and understanding of its application that could materially shorten, and in some instances even eliminate, the painful and unproductive induction period that all too often is the lot of the scientist entering industry.

MILTON HARRIS, PH.D.
Vice President, Gillette Corporation (ret.)
Former Chairman, Board of Directors
American Chemical Society

PREFACE

All of a sudden the organic chemicals industry is middle-aged. Gone are the heady days of the 1950s with new products and processes evolving so fast that improvement of existing processes seemed hardly worthwhile. Gone, too, are the exciting 1960s when every ethylene cracker had twice the capacity of the last one, and it really looked as if the trees might grow to the sky.

Instead, the growth rate has slackened. US natural gas supplies are diminishing. The price of oil has shot up, and "doom boom" proponents point to the rapid depletion of irreplaceable natural resources and speculate on the catastrophe their exhaustion will create. Environmental pressures have caused the chemical industry to take a hard look at its safety procedures and safeguards for workers with toxic chemicals.

The organic chemicals industry has become a large industry playing a large part in the economy. It has matured, and maturity brings the ability to see things in perspective. These two volumes are just such a perspective, presenting a picture of the organic chemical industry in terms of both chemistry and technology. Chemistry may be defined as the creation of molecules; technology as the application of those molecules to the creation of goods and services.

Of course we have written in detail about the petrochemicals industry—it is the dominant side of the industry—but we have also looked at other raw materials that are commonly used and that might be used more if oil or natural gas were not available. We have emphasized the versatility of industrial organic chemistry and tried to outline an alternative technology to which we could turn if the necessity arose.

We appreciate the problems of describing industrial organic chemistry. Tens of thousands of organic chemicals are manufactured. Processes are legion; information on them is frequently confidential or buried obscurely in patents. Small wonder that the average organic chemistry textbook is content to pass lightly over industrial processes.

It is possible to simplify the picture, however, by a study of where industrial organic chemicals come from and what they are used for. Of the 250 billion pounds of chemicals produced annually in the United States, 85 billion pounds are produced from 25 billion pounds of ethylene. Seventy percent of US organic chemicals are based on ethylene, propylene, and benzene. If we add to these three starting materials the C_4 olefins, toluene, xylenes, and methane and recognize that small but important groups of chemicals are derived from coal, fats and oils, and carbohydrates, then the question of where chemicals come from becomes more straightforward.

A study of the uses of organic chemicals, that is, of their technology, is similarly simplified by the realization that over 40% of them find their way into the polymer production and processing industries. There are, of course, many other industries that use chemicals. Their diversity makes organization more difficult, and we have tried to give a broad overall picture together with a study in depth of selected cases.

Furthermore, we have tried to look at organic chemicals not only from the point of view of the producer who is manufacturing a specific chemical entity but also from the point of view of the consumer who is buying not a chemical as such but a property that he requires. We discuss many of the materials downstream of the organic chemicals industry—elastomers, fibers, plastics, adhesives, surface coatings, solvents, plasticizers, lubricating oils, surfactants, drugs, food chemicals, dyes and pigments, and agrochemicals—and have considered why their manufacturers buy organic chemicals and how they use them. We think these two volumes are unique among textbooks in dealing with the problems of formulating chemicals into saleable mixtures.

Finally, we have tried to place the US and world organic chemicals industries in their contexts in US and world economies at the beginning of the last quarter of the twentieth century.

These volumes are suitable for anyone who wishes to gain a perspective on the organic chemicals industry and are intended as a textbook for chemistry and chemical engineering graduate students and for undergraduates who have completed a course in organic chemistry. Part I will be useful for a one-quarter or one semester course. Part II provides a textbook for a second quarter or semester and is a useful reference for a shorter course.

HAROLD A. WITTCOFF
BRYAN G. REUBEN

Beer-Sheva, Israel
London, England
January 1980

ACKNOWLEDGMENTS

We acknowledge with pleasure the important contributions of Dr. Michael Burstall who reviewed the entire book and of Dr. C. Macosko who reviewed Chapter 4. The help of the library staff of General Mills Chemicals, Inc., particularly Mrs. Margaret Drews, Dr. William Mayer, and Dr. Ronald Dueltgen, has been invaluable. Heartfelt gratitude is due Mrs. Shirley Lindell who typed tediously innumerable drafts and cheerfully met deadlines. We thank David Reuben for assistance with proofreading.

H.A.W
B.G.R.

CONTENTS

CONTENTS FOR PART TWO

INDUSTRIAL ORGANIC CHEMICALS IN PERSPECTIVE

PART ONE: RAW MATERIALS AND MANUFACTURE

Chapter 0
HOW TO USE "INDUSTRIAL ORGANIC CHEMICALS IN PERSPECTIVE" PARTS I AND II

0.1 STRUCTURE AND LAYOUT

Part I starts with an overview of the chemical industry and its place in the national economy (Chapter 1). It continues with a discussion basic to the chemistry of industrial organic chemicals. Where do organic chemicals come from? Chapter 2 deals with petroleum sources and Chapter 3 with sources other than petroleum. Chapter 4 describes how polymers are made, and Chapter 5 describes aspects of catalysis that are basic to modern chemistry. In Chapter 6 we speculate about the future of the chemical industry and the role of the chemist in it.

Part II describes the technology of the industry—how chemicals are used. Chapter 1 considers the various industries that buy chemicals and how those chemicals are chosen. Polymers are a major outlet for chemicals. Thus Chapters 2 to 6 cover the uses of polymers in plastics, fibers, elastomers, surface coatings, and adhesives. Chapters 7 to 14 deal with some of the industries using chemicals that are not mainly polymers, notably surface active agents, drugs, solvents, lubricating oils, plasticizers, agrochemicals, dyes and pigments, and food chemicals.

Each chapter is numbered, and these chapters are divided and sub-

1

divided. For example Chapter 3 deals with sources of chemicals other than natural gas and petroleum; Section 3.3 covers carbohydrates, and Section 3.3.2 refers to starch in particular. All cross-references—and we have provided them profusely—refer to the numbered subsections and indicate in which volume these occur. A reference "see note" directs the reader to the notes and references section at the end of each chapter, and the note is found under the decimal heading in which the reference occurs.

0.2 STANDARD INDUSTRIAL CLASSIFICATION

We define all industries and branches of industry according to the US Standard Industrial Classification (SIC), which appears on the Standard Industrial Classification Manual: 1972 available from the US Government Printing Office, Washington, DC 20402. The classification divides the economy into broad sectors, such as the manufacturing industry, which are then subdivided further. Thus the manufacturing industry is split into 20 two-digit major groups such as SIC 28, chemicals. These are sub-divided into industry groups such as SIC 282, plastics and synthetic fibers, and a fourth digit defines a specific industry such as SIC 2823, cellulosic man-made fibers. The industries associated with the Chemical and Allied Products Industries are shown in Table 1.2.

The definition of the chemical industry in countries other than the United States differs in detail from that given above, and those wishing to tackle official statistics should beware of such pitfalls.

0.3 UNITS AND NOMENCLATURE

The widespread introduction of the SI (Système International) system of units based on the meter, the kilogram, and the second has worsened the chaos among the units used by the chemical industry. Three kinds of ton are in common use—the short ton (2000 lb), the metric ton or tonne (1000 kg or 2204.5 lb), and the long ton (2240 lb). US statistics tend to be given in millions of pounds which are unambiguous, and we have followed this example apart from occasional references to tons. For the most part we try to quote figures in the units actually used by industry (e.g., petroleum is measured in barrels, benzene in gallons, and toluene in pounds). Conversions into the SI or other well-known scientific units are included when necessary. A table of conversion factors is given on the end papers.

Similarly, in naming chemicals we tend to use the names conventional in industry rather than the more academic nomenclature of the International Union of Pure and Applied Chemistry (IUPAC). Thus we write hydrogen, not dihydrogen; ethylene, acetylene, and acetic acid, not ethene, ethyne, and ethanoic acid. We will refer to the compound $C_6H_5CH(CH_3)_2$ as cumene, the name by which it is bought and sold, rather than by the more informative names of isopropylbenzene, 2-phenylpropane, or (1-methylethyl)benzene. Similarly, the word ethanal would be likely to be misread as ethanol in industry where the compound would still be known as acetaldehyde. So important is trivial nomenclature that the pharmaceutical industry could not operate without it. We apologize to those who are unacquainted with the trivial names, but we feel that this best serves our aim of introducing the student to chemical industry practice.

0.4 GENERAL BIBLIOGRAPHY

In many ways the greatest service a book like this can provide is to introduce the student to the industrial chemical literature. We follow each chapter with an annotated bibliography that lists some of the standard literature on the subject of the chapter, cites the sources of much of our own information, and. adds occasional notes to matters discussed in the chapter.

Certain books are of general interest and are discussed below. When referred to later in the book, they are cited in an abbreviated form. For example, *The Chemical Economy*, by B. G. Reuben and M. L. Burstall, Longman, London, 1974, is referred to as *The Chemical Economy*.

In our bibliographies we largely confine ourselves to material published after 1970. References to earlier works may be found in *The Chemical Economy*, *Kirk-Othmer*, and other encyclopedias (see below) and in the bibliography, *Literature of Chemical Technology*, J. F. Smith, Ed. Advances in Chemistry Series No. 78, American Chemical society, Washington, DC 1968.

0.4.1 ENCYCLOPEDIAS

The most important single work of reference is Kirk-Othmer's *Encyclopedia of Chemical Technology*, 22 volumes plus one supplementary volume, 2nd ed., Interscience, New York, 1962–1971. It provides comprehensive and well-referenced coverage of almost every aspect of industrial chemistry. The earlier volumes are inevitably dated, but seven volumes of the third edition have already been published, and many more should be available by the time this book appears.

The *Encylopedia of Polymer Science and Technology*, N. Bikales, Ed., 15 volumes, Interscience, New York, 1964–1971, provides comprehensive coverage of polymer chemistry. It is weak on technology but is well-referenced. It is now somewhat dated, and a new edition would be welcome.

There is more of a chemical engineering bias in the *Encyclopedia of Chemical Processing and Design*, J. J. McKetta, Ed. Dekker, New York, 1977. Four volumes of this encyclopedia had appeared at the time of this writing.

A materials approach is adopted in *Materials and Technology: A Systematic Encyclopedia of the Technology of the Materials Used in Industry and Commerce*, eight volumes, Longman, London, 1968–1975. This series is published in the United States under the title *Chemical Technology: An Encyclopedic Treatment*, Barnes and Noble, New York. It is strongest on topics outside the mainstream of the heavy organic and inorganic chemicals industries, and the approach is original and useful.

A single volume work is *Chemical and Process Technology Encyclopedia*, D. M. Considine, Ed., McGraw-Hill, New York, 1974. It is biased towards engineering rather than technology and is poorly indexed and contains insufficient references. Nonetheless it contains much useful material for those who are prepared to search.

The flow sheets for 711 processes with scarcely any explanatory text are gathered in the *Organic Chemical Process Encyclopedia*, M. Sittig, 2nd ed., Noyes Development Corporation, 1969. The diagrams tend to be strong on reactors and weak on separation processes. They are an invaluable adjunct to some other book, but because the flow sheets are taken from the patent literature there is an unfortunate lack of information on which processes have actually been commercialized.

Invaluable to those to whom it is available is the *Chemical Economics Handbook*, Chemical Information Services, Stanford Research Institute, Menlo Park, CA. This 24-volume work contains a wealth of up-to-date information on both the technology and the markets of numerous industrial chemicals. Since new data are provided continually, figures are seldom older than four years, and most are more recent. Because of its expense, this compendium is usually found only in industrial libraries, and the subscriber agrees to keep the information confidential.

0.4.2 BOOKS

A one-volume guide to the technology and economics of the chemical industry is *The Chemical Economy*, B. G. Reuben and M. L. Burstall,

Longman, London, 1974. It provides an overview of the industry, emphasizes the organic chemicals industry, and is biased to some extent toward European practice. It contains annotated bibliographies that supplement those given here.

Two classic books on the chemical industry which have appeared in new editions are *Riegel's Handbook of Industrial Chemistry*, J. A. Kent, Ed., 7th ed., Van Nostrand-Reinhold, New York, 1974, and *Chemical Process Industries*, R. N. Shreve and J. A. Brink Jr., 4th ed., McGraw-Hill New York, 1977. *Riegel* first appeared in 1928 and is now a multiauthor survey of the Chemical and Allied Products Industry. *Chemical Process Industries* is perhaps weak on organic chemicals and is printed in a typeface that makes it difficult to read, but it has unequaled scope and authority.

Faith, Keyes, and Clark's *Industrial Chemicals* first appeared in 1950 and was updated in 1957 and 1965. The appearance of the fourth edition in 1975, revised by F. A. Lowenheim and M. K. Moran, Wiley-Interscience, New York, was a major event. The book provides details of manufacture and markets of the 140 most important chemicals in the United States and can be unreservedly recommended from every aspect except the price.

Survey of Modern Industrial Chemistry, G. A. Cook, Ann Arbor Science, Ann Arbor, MI, 1975, is a shorter book that illustrates the operation of the chemical and allied products industry by examples selected from the different branches of the industry.

Chemistry in the Economy, M. Harris and M. Tischler, Eds., American Chemical Society, Washington, DC, 1973, provides, as its title indicates, an excellent overview of the impact of chemical technology on the economy by describing how chemistry is used in numerous industries. It also gives excellent insight into the morphology of the chemical industry.

Chemicals are discussed from the point of view of the consumer in an interesting and original book *Chemistry in the Market-Place.* Ben Selinger, 2nd ed., Australian National University Press, Canberra, 1978. The formulation of many domestic products is described together with the reasons for the various additives and the theory behind them.

Among the books dealing specifically with the organic chemicals industry is *An Introduction to Industrial Organic Chemistry*, P. Wiseman, Halsted Press, New York, 1972. It is well-organized and well-written and is oriented toward the pure chemistry that provides a base for technology. A second edition appeared in 1979.

Basic Organic Chemistry V: Industrial Products, J. M. Tedder, A. Nechvatel, and A. H. Jubb, Wiley, Chichester, 1975, is the fifth volume in a series on organic chemistry, but the title is somewhat misleading since it can stand by itself as a textbook on industrial organic chemistry. It is a multiauthor survey biased toward chemistry rather than technology and toward British practice. Insufficient references are given Although the book looks dull it is a mine of valuable information for the specialized reader.

Principles of Industrial Chemistry, C. A. Clausen III and G. Mattson, Wiley-Interscience, New York, 1978, is aimed at chemists and provides an enthusiastic introduction to chemical process principles, process development, and various commercial aspects of the chemical industry.

The Structure of the Chemical Processing Industries, J. Wei, T. W. F. Russell and M. W. Swartzlander, McGraw-Hill, New York, 1979, is a book similar in structure to *The Chemical Economy*, but it deals with the economic structure of the chemical industry in much greater depth and spends less time on chemistry and technology.

There are two valuable books in German, *Der Absatz in der Chemischen Industrie*, H. Kölbel and J. Schnitze, Springer, Berlin/New York, 1970, and *Industrielle Organische Chimie*, K. Weissermehl and H. J. Arpe, Verlag chimie, Weinheim, 1976. The latter has been translated into English by A. Mullen under the title of *Industrial Organic Chemistry*.

0.4.3 AUDIOVISUAL AIDS AND CASE STUDIES

In the last few years the American Chemical Society in the United States and a number of organizations in the United Kingdom have started to take an interest in the teaching of industrial chemistry, and a number of aids to learning have become available. Of greatest interest are the numerous audio courses produced by the American Chemical Society (1155 16th St., NW, Washington, DC 20036). In particular, *Industrial Organic Chemistry*, H. Wittcoff, 1979, should interest readers of this book. Many US chemical companies have films they will loan, but there is no central catalog of these, and each company must be contacted separately.

The *AAAS Science Film Catalog*, American Association for the Advancement of Science, Washington, D. C. and R. R. Bowker, New York/London, 1975, lists a large number of science films, but most appear to be on an elementary level.

Probably the best set of industrial chemistry case studies in existence was produced by G. Mattson and C. A. Clausen at the Florida Technological University, Orlando. Topics covered are chloromycetin, styrene, urea, polyurethanes, evaluation of crude oil, and the development of a partial replacement for zinc in zinc-rich coatings. Cyclostyled copies were distributed through *CHEMTECH* in 1975, but they are no longer available. The authors indicate they are planning to publish them as a book, and the study of urea has already appeared in *Principles of Industrial Chemistry*, listed above. The Chemical Society, Burlington House, Piccadilly, London WIV OBN publishes a cyclostyled case study on natural gas *What Happens when the Gas Runs Out . . .* ? by F. Percival and A. H. Johnstone 1977. *Intercollegiate Bibliography 1974*, *Selected Cases in Administration*, Intercollegiate Case Clearing House, Soldiers Field, Boston MA 02163 lists many case studies some of which (e.g., Industrial Chemicals Inc., Mobil Chemical Co., Reichold Chemicals Inc.) deal with the chemical industry.

Film catalogs in the United Kingdom tend to date rapidly and have only a tiny section devoted to industrial chemical films. In 1970 the Royal Institute of Chemistry produced *The Index of Chemistry Films* available from the Chemical Society Distribution Centre, Blackhorse Rd., Letchworth, Herts, England which recorded the material then available. The Education Foundation for Visual Aids, 33 Queen Anne St., London W1M OAL, produces an eight-volume catalog with industrial chemistry in volume 7, but it badly needs updating. More recent is *The Central Film Library Catalogue 1977* available from the Central Film Library, Government Building, Bromyard Ave., Acton, London W3 7JB. Large companies such as Imperial Chemical Industries, Unilever, and British Petroleum produce their own films, but most of them are publicity oriented and are rarely suitable for university education.

The Open University in the United Kingdom is an organization that teaches by means of television and radio broadcasts backed up by tutorial and course work and summer schools. Audio and audiovisual materials and supporting printed texts may be purchased from Open University Educational Enterprises Ltd., 12 Cofferidge Close, Stoney Stratford, Milton Keynes MK11 1DY, England. Video tapes include the *Industrial Preparation of Acetic Acid* (the UK route from naphtha is described), *Ethylene Production*, and a case study of *Poly(vinyl chloride)*. Accompanying books include *Technology Foundation Course Unit 25, Chemical Processes*, R. Harrison and G. Weaver, and the *PVC Case Study*. For anyone with an interest in the development of the chemical industry there are some excellent video tapes from the course on Science and the Rise of Technology since 1800 including a rare film of the operation of the Leblanc process. There are audio tapes on *Pasteur*, *'Paraffin' Young*, and *From Dyes to Drugs*, and a book on *The New Chemical Industry*, C. Russell and F. Greenaway, 1973.

A series of sets of 12 slides plus notes on *Modern Industrial Chemistry* is produced by N. Hunter Filmstrips, Mutton Yard, 46 Richmond Road, Oxford OX1 2JT, England. Compiled by T. E. Rogers, they deal mainly with inorganic chemicals, but there is currently a set on petroleum refining.

The Joint Matriculation Board, Manchester M15 6EU, England, has produced an excellent book of *Chemistry Case Studies* including one comparing five routes to phenol. The flow sheets for the processes discussed are available as sets of transparencies and overlays for use with an overhead projector from Phillip Harris Biological Ltd., Oldmixon, Weston-super-Mare, Avon BS24 9BJ, England.

0.4.4 JOURNALS

The serious student of the chemical industry should follow the trade press. New products and processes, changes in the structure and prospects of the industry, and so on take several years to get into books whereas the magazines cover them immediately.

International journals are reviewed in *The Chemical Economy*. A general selection for English-speaking readers would be *CHEMTECH* [monthly, American Chemical Society (ACS), Washington]; *Chemical and Engineering News* (fortnightly, ACS, Washington); *Hydrocarbon Processing* (monthly, Gulf Publishing, Houston, TX); *European Chemical News* (weekly, IPC International Press, London); and *Chemical Age International* (weekly, Morgan Grampian, London). Prices of most industrial chemicals are to be found in *European Chemical News* and *The Chemical Marketing Reporter* (weekly, Schnell Publishing Co., New York).

0.4.5 PATENTS

Patents are a device whereby the government grants an inventor the sole right to exploit his invention for a period of 17 years in the United States, 20 years in the United Kingdom, and similar periods in other countries. In return, the inventor discloses details of the discovery or invention in his patent specification.

Incidentally but importantly, patent specifications are also a source of technical information. They often disclose information at a much earlier date than the scientific literature; sometimes they are the only source of such information. Negative results often appear in patents but not in scientific journals, and knowledge of what has been tried without success may save the working scientist much time.

Academic scientists shun patents because the introductions and claims are written in legal jargon with its long convoluted sentences. Libraries shun them because they are published as individual items and are difficult to collect and bind. They have, however, one overwhelming advantage.

They are classified by subject and can be subscribed to in this way. A copy of a US patent costs a mere 50 cents and of a UK patent 95p.

Patent applications are numbered consecutively as they are received by the US Patent Office (US serial number), and when the patent is granted, it is assigned another number (US patent number). Other patent offices are similar.

Brief accounts of patents appear in the chemical trade literature, *Chemical Abstracts* publishes a numerical patent index that lists each patent number together with its corresponding *Chemical Abstracts* abstract number, country of origin, and serial order. It also provides a worldwide list of major patent offices and their addresses. *Chemisches Zentralblatt* (Akademie Verlag, Berlin) offers a similar service together with a guide to its use, *Chemisches Zentralblatt: das System*. Derwent Publications Ltd., Rochdale House, Theobald's Road, London WC1, England, publishes analyses and abridgements of patents from every country, classified by subject, and provides monthly bulletins, for example, *Organic Patents Bulletin* and *Pharmaceutical Patents Bulletin*.

The *Official Gazette*, copies of patents, coupon books (a convenient way to pay for copies), listings of patents by subject, copies of foreign patents, and much other information may be obtained from the Commissioner of Patents and Trademarks, Washington DC 20231. Though many official and commercial organizations exist to help the student of the patent literature, a thorough search can be conducted only at the National Patent Office Library, Washington DC.

In the United Kingdom the equivalent of the *Official Gazette* is the *Official Journal (Patents)*, and it and other information is available from the Patents Office, St. Mary Cray, Orpington, Kent BR5 3RD, England. A thorough search can be carried out at the Patent Office Library, 25 Southampton Buildings, High Holborn, London WC2A 1AW.

Information on subject codes and many other aids to patent searching may be found in *Kirk-Othmer's* encyclopedia and in *Patents—A Source of Technical Information*, Central Office of Information, HMSO London 1975.

0.4.6 TEACHING PROGRAMS

Articles on the teaching of industrial chemistry include "Objectives in the Teaching of Chemistry," M. L. Burstall and B. G. Reuben, *Chemistry and Industry*, 1794 (1968), "Humane Education for the Industrial Chemist," B. G. Reuben, *New Scientist*, 5 November 1970, p. 282; "The Professional Chemist in Industry," J. P. Kennedy, *CHEMTECH*, **4,** 156 (1974); "Cases in Chemical Technology" C. A. Clausen and G. Mattson,

CHEMTECH, **5,** 535 (1975); "Surrey's Course in Chemical Technology," B. G. Reuben, *CHEMTECH*, **5,** 402 (1975); "How It's Really done," I and II, H. Wittcoff, *CHEMTECH*, **7,** December 1977, p. 754 and **8,** April 1978, p. 238; and "An Experimental Course in Industrial Chemistry, H. Wittcoff, *Journal of Chemical Education*, **52,** September 1975, p. 596.

Chapter One
THE CHEMICAL INDUSTRY

1.1 THE NATIONAL ECONOMY

The United States, Western Europe, and Japan are the most complex societies that have ever existed. Division of labor has been carried to the point where almost everyone performs a highly specialized task and relies on many others to provide him with the goods and services he needs. In return for these goods and services he provides his output to satisfy the needs of others. All men are brothers in a material sense just as they should be in a moral sense.

The interdependence of a society's activities may be seen more clearly if its economy is divided into specific industries or groups of industries This is normally done according to the standard industrial classification of the Bureau of the Census in which each industry is allocated a code number. Table 1.1 shows the main sectors of a developed economy. These sectors are interdependent. For example, the manufacturing industry may draw heavily on the output of the mining sector by buying iron ore from which to make steel. In turn it may convert that steel to machinery to sell back to the mining industry who will use it in mining operations. The economy as a whole and the various sectors separately

Table 1.1 Main Sectors of a Developed Economy[a]

The Economy	Classification	Value[b] added US 1975 (billion $)
Agriculture, forestry and fisheries	(01–09)	54.8
Mining	(10–14)	37.6
Construction	(15–17)·	66.5
Manufacturing	(20–39)	346.0
Transportation, communication, and electric, gas, and sanitary services	(40–49)	132.2
Wholesale trade	(50–51) ⎫	
Retail trade	(52–59) ⎭	272.4
Finance, insurance, and real estate	(60–67)	209.4
Other services, including medicine, education, social services, and entertainment	(70–97)	181.8
Government and Government enterprises	(98)	200.6
Nonclassifiable establishments	(99)	15.0
		1516.3

[a] Based on US Bureau of the Census, Annual Survey of Manufactures, 1975.
[b] Value added is value of shipments less cost of materials, supplies, containers, fuels, purchased electricity and contract work plus net change in finished goods and work-in-progress inventory and value added in merchandising activities of manufacturing establishments. It is thus the value added to all the inanimate inputs to an industry by the people working in it. The total value added throughout the economy is the gross national product, the sum of wealth produced by the nation, in this case $1516.3 billion.

require various inputs in order to function. These are usually described as the resources of production—labor, land, and capital. Labor is people and their beliefs—the economic and political system and the education and ethical philosophy of its citizens. Land is not only real estate but also the natural resources found in a country, for example, iron ore, natural gas, oil, fresh water, and fresh air. Capital is property of all kinds used in production such as buildings or machine tools. Most economies lack certain inputs and are forced to import them. Oil and metal ores such as bauxite are examples.

The economy converts these inputs into outputs—the goods and services demanded by people who live in the society, the so-called final consumers. It also exports some of the outputs to pay for the imports.

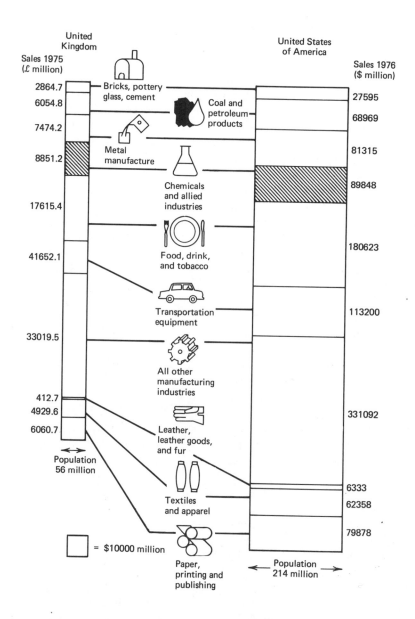

Figure 1.1 Rank of chemicals and allied products industry among manufacturing industries—manufacturers' shipments in the United States and the United Kingdom, 1975. Width of band is proportional to population, and area of band is proportional to value of shipment; hence length of band is proportional to shipment per head of population. The figure for US coal and petroleum products does not include foreign operations of US companies. Total shipments are about $124 billion. Transportation equipment represents about 70% motor vehicles in the United States and 62% in the United Kingdom.

Foreign trade is important even for large, almost self-sufficient countries like the United States, and smaller and poorer countries are often dependent on it to be able to eat. The United Kingdom, for example, produces only a small proportion of the food it consumes and has to export its manufactured goods in order to import food. That this was so even in the days of Britain's imperial greatness shows that foreign trade is part of an international division of labor. It often is more economical for a country to import certain of its needs and to pay for these with exports rather than to try to satisfy them itself.

The United States is better off importing bauxite (aluminum ore) from Surinam and selling silicon "chips" for microcircuits than she would be without foreign trading, and the exporting countries benefit similarly. It would not be reasonable for them to try to master the high technology involved in silicon "chips"; neither would it be reasonable for the United States to try to supply all the aluminum she needs by working lean deposits of bauxite.

The inputs to an economy are in a sense its limitations. Resources are always available in restricted quantities; and even those that seem limitless now may be depleted in the future.

The outputs from an economy, at least in a free society, are the result of the needs and demands of the people who have discretionary income and make decisions about how to use it.

Certain of these outputs fulfill basic material needs—food, clothing, housing, minimal health services, and security. Possibly transportation and certain facilities for recreation should also be included. As a nation becomes richer, the demand arises for more costly food, clothing, housing, and health services, but these still represent the satisfaction of material needs. At a level beyond material needs is the desire for emotional satisfactions and for intellectual and ethical as well as physical fulfillment.

The chemical industry is a subdivision of the manufacturing industry (Fig. 1.1) but because of its nature and because of the underlying interdependence of all industries it contributes to almost every one of the various outlets. For example foods—carbohydrates, proteins, and fats—are chemicals. Their processing, preservation, and packaging depend in large part on chemical principles. Clothing is made from natural, semisynthetic, and man-made fibers. Even natural fibers require dyes, sizes, and other chemicals for their processing. In the United States about 60% of all fibers used are synthetic. This is fortunate because the demand for textiles in developed countries has far outstripped the availability of land for growing cotton or flax.

Housing too depends on chemicals ranging from the simple inorganics used in concrete and glass to the complicated organics used in engineering plastics, coatings, and wood preservatives. The modern automobile could

not exist without the chemists' materials, and the average US car produced in 1976 contained 160 lb of plastics, not to mention the synthetic
rubber in the tires and shock absorbers. The figure will probably double
in a few years. More exotic forms of transportation such as rockets and
space shuttles are even more dependent on the chemical industry.

The main improvements in health in the last half century have resulted
from chemotherapeutics, and bacterial diseases such as tuberculosis and
bacterial pneumonia have been virtually eliminated in developed countries by chemicals known as antibiotics. Cancer is still unconquered, but
the most hopeful strides have been based on chemicals.

Recreation too has been influenced by modern chemistry. The skier
benefits from epoxy-glass fiber skis and polyurethane boots, while the
mountain climber is safer and more comfortable because of his nylon or
polyester ropes and synthetic fiber clothes.

The relationship of human fulfillment to technology is complex and
begs the question "What is happiness?" Nonetheless chemistry's contribution to modern technology is also a contribution to the possibility of
fulfillment, for modern technology gives men and women unprecedented
amounts of free time in which to "do their own thing." Whatever
fulfillment may involve, it is scarcely compatible with a life of subsistence
farming and an infant mortality rate of 200 per 1000 live births, which
was the lot of most of our ancestors.

That chemistry has also contributed to modern warfare and that atomic
weapons make the total annihilation of mankind a distinct possibility
shows, however, that the use of technology (which includes chemistry)
requires value judgments. Modern drugs can be used to aid the mentally
sick or to render the mentally well insane. Atomic power can be used for
peaceful or violent purposes. Steel can be used for swords or
ploughshares, and paleolithic stone axes could be used to split heads as
easily as to split wood.

Modern technology will not go away, and if it did few of us would be
better off by any conceivable standards. What we must hope for is that
wise decision making will accompany technological achievement. An
understanding of the role of chemistry (and other technologies) in our
society is a contribution toward this aim.

1.2 SIZE OF THE CHEMICAL INDUSTRY

The chemical industry is ubiquitous, but it is only one sector of the
manufacturing industry among many. In this and subsequent sections we
try to illustrate its size, its ramifications, and its place in the national
economy.

The gross national product of the United States in 1976 was $1.69
trillion. To this total the manufacturing industry contributed about 25%.

Table 1.2 Chemicals and Allied Products Industry (SIC 28) 1975[a]

SIC Number		Industry	Value of Product Shipments (million $)
281		Industrial Chemicals	
	2812	Alkalies and chlorine	1,630
	2813	Industrial gases	919
	2816	Inorganic pigments (except carbon black)	972
	2818	Industrial organic chemicals n.e.c.[b]	13,200
	2819	Industrial inorganic chemicals n.e.c.[b]	4,657
282		Plastic materials and synthetics	
	2821	Plastics materials and resins	7,899
	2822	Synthetic elastomers	1,805
	2823	Cellulosic man-made fibers	780
	2824	Organic fibers, non-cellulosics	4,042
283		Drugs	
	2831	Biological products	758
	2833	Medicinals and botanicals	1,484
	2834	Pharmaceutical preparations	8,178
284		Soap, cleaners and toilet goods	
	2841	Soap and other detergents	3,215
	2842	Polishes and sanitation goods	2,189
	2843	Surface-active agents	961
	2844	Toilet preparations	5,150
2851		Paint and allied products	4,678
2861		Gum and wood chemicals	363
2865		Cyclic intermediates and crudes	4,260
2869		Industrial organic chemicals	14,017
287		Agricultural chemicals	
	2873	Nitrogeneous fertilizers	2,836
	2874	Phosphoric fertilizers	2,466
	28752	Mixing of fertilizers	1,285
	2879	Agricultural chemicals n.e.c.[b]	2,266
289		Miscellaneous chemical products	
	2891	Adhesives and sealants	1,471
	2892	Explosives	412
	2893	Printing ink	691
	2895	Carbon black	307
	2899	Chemical preparations n.e.c.[b]	2,951

[a] Derived from US Bureau of the Census, Annual Survey of Manufactures, 1975.
[b] n.e.c. = Not elsewhere classified, that is the industrial organic chemicals category excludes the industrial organic materials already counted in sections 2812–2816

Table 1.3 Projected Growth of the Chemical and Allied Products Industry in the USA (Billion $)

	Dollar Volume 1976 (est.)	Projected Dollar volume 1985
Chemical industry	48.6	94.4
Allied products industry	52.8	120.6
	101.4	215.0

The largest manufacturing industry is food and kindred products followed by transportation equipment and mechanical engineering and the other industries shown in Figure 1.1 In terms of value added (see footnote to Table 1.1) the chemicals and allied products industry ranks third among manufacturing industries both in the United States and in the United Kingdom, though this is not apparent from Figure 1.1, which lists shipments—not value added. If foreign operations of US petroleum companies are included, then the chemical and allied products industry ranks only fourth among US industries.

The chemicals and allied products industry is subdivided in the standard industrial classification into many component industries. Broadly speaking, the chemical industry (SIC 281) isolates or synthesizes chemicals, whereas the allied products industries (SIC 282–289) modify, formulate, and package products based on those chemicals.

The division by shipments is shown in Table 1.2 and illustrated in Figure 1.2, which also shows the division by value added. The industry grossed about $100 million in the United States, in 1976. Indicated in Table 1.3 is our projection, assuming no major economic disruptions, for the size of the industry by 1985 in constant dollars, that is dollars with 1976 buying power. If inflation were taken into consideration the projections would be much higher.

Comparison of the shipments and value added diagrams shows that so-called "fine chemicals," such as drugs, dyestuffs, and toilet preparations make a larger contribution to the chemical industry's value added than they do to its shipments. They tend to be high-priced products with specialized markets, and their manufacture is less capital and more labor intensive than the manufacture of the run-of-the-mill general chemicals. Their importance to the chemical industry is best represented by the value added figure, which, for example, indicates the importance of the pharmaceutical sector.

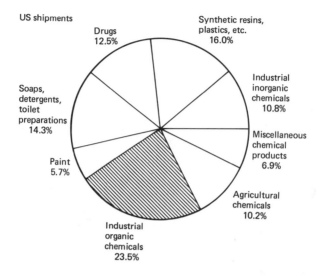

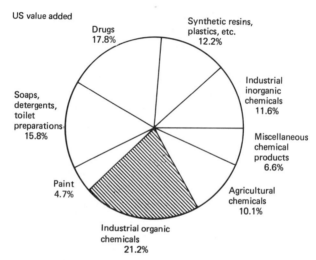

Figure 1.2 Subdivision of the chemical and allied products industry in the USA and UK by value of shipments and value added.

UK shipments

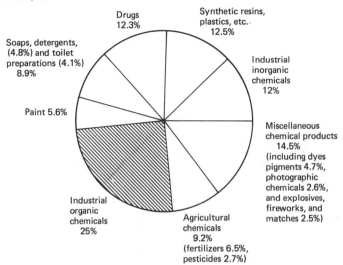

Drugs 12.3%

Synthetic resins, plastics, etc. 12.5%

Soaps, detergents, (4.8%) and toilet preparations (4.1%) 8.9%

Industrial inorganic chemicals 12%

Paint 5.6%

Miscellaneous chemical products 14.5% (including dyes pigments 4.7%, photographic chemicals 2.6%, and explosives, fireworks, and matches 2.5%)

Industrial organic chemicals 25%

Agricultural chemicals 9.2% (fertilizers 6.5%, pesticides 2.7%)

UK value added

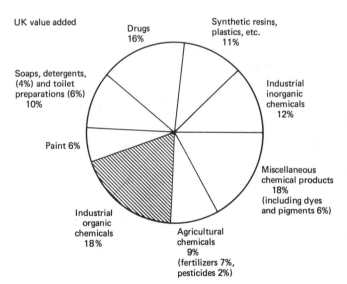

Drugs 16%

Synthetic resins, plastics, etc. 11%

Soaps, detergents, (4%) and toilet preparations (6%) 10%

Industrial inorganic chemicals 12%

Paint 6%

Miscellaneous chemical products 18% (including dyes and pigments 6%)

Industrial organic chemicals 18%

Agricultural chemicals 9% (fertilizers 7%, pesticides 2%)

Figure 1.2 (*Continued*)

1.3 CHARACTERISTICS OF THE CHEMICAL INDUSTRY

The chemical and allied products industry has certain well-defined characteristics, listed in Table 1.4, that govern its attitudes and its performance. We shall discuss them in turn.

1.3.1 GROWTH

The chemicals and allied products industry in most developed countries has grown over the last quarter century at approximately twice the rate of the manufacturing industry as a whole. In the United States the industry grew at 8.5–9% per year from 1954 to 1974 and in the United Kingdom at 5.8%. Other data are given in Table 1.5 for the years 1964–1974. There was a severe recession in 1975, but the industry rebounded in 1976, and in the United States there was an increase in actual dollars of about 17%. Between 5 and 8%, however, was due to inflation.

Economists' opinions are divided on the future growth of the industry. On one hand it is possible that it is approaching maturity. The petrochemical revolution has taken place; there are no dramatic breakthroughs in prospect. Sooner or later the industry must slow down and grow at the same rate as the economy as a whole. The industry faces many problems. The price of oil has risen rapidly and is likely to continue rising, whereas it dropped steadily in real terms between 1950 and 1971. Problems of pollution, worker safety, and depletion of natural resources are likely to occupy time and effort and consume research expenditures and capital investment that might otherwise have been devoted to "growth." In the United States the Food and Drug Administration (FDA) and similar regulatory organizations have tightened their regulations, and this has been reflected in a dramatic drop in approved new pharmceuti-

Table 1.4 Characteristics of the Chemical Industry

1. Rapid growth
2. High research and development (R&D) expenditures
3. Intense competition
4. Capital intensity and economies of scale
5. Rapid obsolescence of facilities
6. Freedom of market entry
7. "Feast or famine"
8. International trade
9. Criticality and pervasiveness

Table 1.5   Growth in the Chemical Industry
for 1964–1974

	Percent per year
Japan	11.7
Western Europe	9.7
United States	8.0
United Kingdom	5.7

cals and agrochemicals. The Toxic Chemical Substances Act may inhibit the introduction of new chemicals generally. The petroleum producing countries are eager to build their own chemical industries, and this will lead to overcapacity. However one looks at it, chemical industry growth in developed countries is bound to slow down.

On the other hand the worries of 1975 might be just a "blip" on the growth curve. The developing countries are unlikely to be serious competitiors for years to come. Public reaction to the banning of saccharin and cyclamates may persude the regulatory bodies to regulate less. The "creativity" of industrial chemists could keep the new processes rolling and overcome the threats of resource depletion.

Probably neither point of view will prove accurate. The future certainly has some surprises in store. One thing is certain—today's students will be making a contribution in one way or another to whichever scenario comes to pass.

1.3.2   RESEARCH AND DEVELOPMENT

The chemical industry is research intensive. It hires many graduates, and most of them work in research and development (R & D) laboratories (see notes). Research expenditure by the world's top chemical companies is shown in Tables 1.6 and 1.10. Three to four percent of sales spent on research is considered normal, but not all chemical companies conform. Companies making heavy inorganic chemicals or fertilizers where technology changes little, if at all, spend less; pharmaceutical companies and specialty chemical companies spend more. Table 1.6 lists companies with high research expenditures. Nearly all are pharmaceutical companies. UK, Japanese, and Italian companies appear to be spending less on research, both as a proportion of sales and in absolute terms, than do Swiss and West German companies.

Table 1.7 shows how research expenditure in the US chemical industry compares with that in other industries. Three other industries, aerospace, electric machinery and communications, and machinery spent more on research in 1976. The data also show that most research is devoted to product improvement and not to new products and processes. Twelve percent was spent on energy-related R&D, and this figure is likely to vary sharply as energy costs and government policies fluctuate. Probably it will consume an increasing percentage of the total.

Research is a risky business, and, since billions of dollars are spent, there has also been research on research to try to find the conditions that maximize the cost effectiveness of an R & D budget. Broadly speaking, there are two extreme attitudes that meet in the middle—technology push and demand pull. In the former the manufacturer discovers the technology and tries to create a market for it; in the latter he examines what the market wants and tries to discover technology to provide it. Television, sulfonamides, and lasers are products of the former approach, for there was no established market for any of these before they were discovered. Hard water detergents, jumbo jets, and automobiles with low exhaust emission are examples of the latter, the mission oriented approach. Demand pull is popular at present, but this is merely a reflection of the maturity of our present technology. Both kinds of research should be part of any large company's game plan.

A logical consequence of the research intensity of the chemical industry

Table 1.6 Research Expenditures for 1976

Rank Order	Top R&D Spenders (% of sales)	
1	Hoffmann, La Roche (Switzerland)	14.8
2	Boehringer Ingelheim (W. Germany)	11.9
3	Upjohn (US)	9.0
4	Sandoz (Switzerland)	9.0
5	Syntex (US)	8.6
6	Eli Lilly (US)	8.4
7	Merck & Co. (US)	8.2
8	SmithKline (US)	8.1
9	Wellcome Foundation (UK)	7.9
10	Ciba-Geigy (Switzerland)	7.6
11	Schering (W. Germany)	7.6
12	G. D. Searle (US)	6.9
13	International Flavors & Fragrances (US)	6.2

Table 1.7 Industrial R&D Expenditures (McGraw-Hill Survey)

	1976 (Billion $)
Total	25.8
By type	
Product improvement	13.44 (52%)
New product development	8.53 (33%)
New process development	3.88 (15%)
	(100%)
Energy-related R&D	3.0 (12%)
By industry group	
Aerospace	6.40
Electric machinery and communications	6.08
Machinery	2.98
Chemical	**2.85**
Automobiles, trucks, and parts	2.37
Instruments	1.56
Oil	0.69
Food	0.39
Rubber	0.31
Fabricated metals	0.36
Paper	0.24
Stone, clay, and glass	0.23
Steel	0.20
Nonferrous metals	0.21
Projected total expenditure for 1979	31.5
Increase over 1976	22%

is its technical complexity. This is illustrated both by the complex reactions described in Chapter 2 based on homogeneous and heterogeneous catalysis and by the fact that modern chemical industry is confined to advanced countries with the necessary social, economic, and intellectual infrastructures. Of the world's 100 largest chemical companies in 1976, 55 were US companies, and another 42 were divided between Japan (12), West Germany (10), United Kingdom ($5\frac{1}{2}$*), France (5), Italy (4), Switzerland (3), and The Netherlands ($2\frac{1}{2}$*). The state-owned chemical companies of countries like Russia with centrally planned economies were not

* Royal Dutch Shell is a joint UK/Holland venture.

included in this survey. Though many of them are strong in heavy inorganic chemicals there is little evidence that they have made much contribution to the more complex petrochemicals field. In pharmaceuticals there is even a gap between the countries of the developed world. The United States, West Germany, the United Kingdom, and Switzerland (and foreign subsidiaries of companies based in these countries) dominate the world drug market.

1.3.3 COMPETITION

Trade in chemicals is international. Although many large companies appear to be virtual monopolies in their own countries (e.g., Imperial Chemical Industries in the United Kingdom), they are generally forced to compete with overseas companies both in their own and foreign markets. This competition arises from a number of factors. First, economies of scale in the chemical industry are large. Companies prefer to build bigger plants than they need and to export the surplus. As their home market grows they export less and eventually may import material until the time is ripe for a new plant. Many countries that lack indigenous chemical industries will be long-term importers of chemicals and provide a battle-ground for exporting companies.

Second, competition occurs because there are frequently several different processes and possible raw materials for the manufacture of a given chemical. As we show there are at least six industrial routes to acrylic acid (Sec. 2.5.1) and five to phenol (Sec. 2.7.1). Acetaldehyde was traditionally made by hydration of ethylene to ethanol followed by oxidation of the latter. Modern technology has provided a one-step process. Most of the acetaldehyde produced was oxidized to acetic acid until processes were discovered that gave acetic acid either in a single step from a hydrocarbon feedstock, thus eliminating three stages from the traditional route, or by an important reaction involving the combination of methanol and carbon monoxide (Sec. 2.4.3). This list could be extended considerably. If a company discovers a new, more economic route to a chemical it will patent it and try to sell the product (and/or license the process) all over the world. In pharmaceuticals, in particular, competition often takes the form of competitive innovation rather than the conventional commercially based competition where price, specifications, technical service, and reliability of delivery are the decisive factors. The chemical industry by any standards is characterized by a high degree of ingenuity. This is reflected in the number of competing processes and by the intense competition resulting from them.

The third reason for competition, which follows from the second, is that a single producer rarely dominates the market. Patent monopolies, although designed to protect the innovator, rarely do for a long period. Development work, especially in pharmaceuticals, often consumes a substantial portion of the time for which a patent is valid. Continuing research rapidly renders older processes obsolescent, and new producers with improved processes seem continually to be entering the market.

1.3.4 CAPITAL INTENSITY AND ECONOMIES OF SCALE

Not only is the chemical industry research intensive, but it is also capital intensive. It produces huge quantities of homogeneous materials, frequently liquids or gases, which can be manufactured, processed, and shipped most economically on a large scale. This was less so through the nineteenth century until World War II. The early chemical industry used more general purpose equipment and operated batch processes that required little capital investment but high labor costs. Typical of such processes were the Leblanc route to sodium carbonate and the benzenesulfonate route to phenol (Sec. 2.7.1).

The petroleum industry was the first to convert to continuous operation on a large scale. The engineering developed for the petroleum industry was applied to the chemical industry after World War II. Plant sizes escalated as dramatic economies of scale became possible. The capacity of a typical ethylene cracker rose from 70 million lb year^{-1} in 1952 to 1 billion lb year^{-1} in 1972. There are slightly larger plants today. Currently there are few batch processes of any size in operation, and substantial economies of scale have become a characteristic of the modern petrochemical industry (Sec. 6.1.3).

Economies of scale arise not only from improved technology but also from purely geometric factors. The capacity of a great deal of chemical equipment (e.g., storage tanks and distillation columns) varies with its volume, that is the cube of its linear dimensions. The cost, on the other hand, is the cost of a surface to enclose the volume and varies with the square of the linear dimensions. Consequently, cost is proportional to (capacity)$^{\frac{2}{3}}$. This is called the square-cube law. It does not apply to all equipment. The capacity of a heat exchanger depends on its surface area so cost is proportional to (capacity)^1 and there are no economies of scale. Control systems are not affected by capacity at all, so cost is proportional to (capacity)^0 and economies are infinite. It is claimed that for a modern petrochemical plant overall, cost is proportional to (capacity)$^{0.6}$.

The size and complexity of a modern chemical plant demand high capital investment. Although other industries invest more capital per dollar of sales, the chemical industry has the highest investment of current

Table 1.8 Business Ratios for Selected Chemical and Nonchemical Companies for 1977 [a]

Company	Industrial Sector	Assets/ Employee ($/person) (000)	Sales/ Employee ($/person) (000)	Sales/ Assets
Standard Oil of California	Oil	387	546	1.41
Proctor and Gamble [b]	Soap and detergents	84	136	1.62
Merck & Co.	Pharmaceuticals	71	61	0.87
Union Carbide	General chemicals	65	62	0.95
IBM	Computers	61	58	0.96
DuPont	General chemicals	57	72	1.27
Bethlehem Steel	Steel	52	57	1.10
Eastman Kodak	Fine chemicals	48	48	1.01
CBS	Broadcasting, etc.	47	82	1.76
Avon Products	Toilet preparations	38	60	1.59
General Motors	Transportation equipment	33	69	2.06
General Mills	Food	23	47	2.01
Brown Group (St. Louis)	Leather	17	37	2.24
Consolidated Foods	Food	16	39	2.43
Genesco	Apparel	11	28	2.58

[a] Based on *Fortune's Top 500* for 1977.
[b] Procter and Gamble is much more capital intensive than most soap and detergent companies and its position on this list is not typical of them.

capital. That means that the chemical industry invests more each year than do such other capital intensive industries as mining, where equipment once bought remains in service for many years.

Capital intensity has a number of corollaries. The return on capital is relatively low. Because high capital investment reduces the labor force required, manpower productivity (i.e., sales per employee) is high. Salaries contribute relatively little to costs (of the order of 20%), and employers need worry less about pay increases than in labor intensive industries such as food or apparel. Consequently, labor relations are unusually good.

The assets of a company are the estimated value of the plant, land, and other capital goods it owns. Such ratios as assets per employee, sales per dollar assets, and sales per employee are measures both of the capital and labor intensity of an industry. Table 1.8 gives the values of these ratios for a range of US companies. For industries as a whole, in 1977 the petroleum refining industry had easily the highest assets per employee and sales per employee. The chemical industry came sixth in the league table for both ratios out of a total of 28 industries.

1.3.5 OBSOLESCENCE OF FACILITIES

As noted, chemical plants have a shorter life than does the equipment used in, for example, mining. This is partly because it is often subjected to corrosive environments. Primarily it is because the pace of technological change has replaced old processes by new ones and has increased the economic size of plants. Until 1973 the chemical industry was certainly characterized by this rapid obsolescence of facilities. Since 1973 the picture has apparently changed. Economies of scale seem to be leveling off (Sec. 6.1.3) so that the incentive to build bigger plants at the expense of the smaller ones has decreased. The cost of new plants has risen steeply. Thus it is worthwhile renovating old ones rather than rebuilding from scratch. It is possible, therefore, that obsolescence of facilities will be slower in the future.

We should point out that rapid obsolescence is not a characteristic of all chemical plants. Equipment for heavy inorganic chemicals where the technology is constant may last for 20 or 30 years. Plants for ammonia and sulfuric acid dating back to the 1930s are said to be operational still.

1.3.6 FREEDOM OF MARKET ENTRY

Another characteristic of the chemical industry is freedom of market entry. Anyone who wants to manufacture bulk chemicals may do so by buying so-called "turnkey" plants from chemical engineering companies. Such companies have processes for preparation of virtually any common chemical and will build a plant guaranteed to operate for anyone who wishes to invest the money. This was the way that many of the petroleum companies gained entry to the petrochemical business and also the way that countries such as the USSR, Spain, Brazil, and Saudi Arabia hope to lay the foundations of their own chemical industries.

A snag, of course, is the entry fee, which has risen sharply in recent years. Without a few hundred million dollars to invest, one would be ill-advised to start manufacturing chemicals on a large scale. In the pharmaceutical industry the cost of developing a new pharmaceutical from discovery to launch in 1976 was estimated at between $15 and $21 million. In the profitable detergent market where capital investment required is much lower there have been few threats to the dominance of Proctor and Gamble, Colgate, and Lever in the United States, or to Proctor and Gamble and Unilever in the United Kingdom. Entry into the chemicals market is free in the sense that there are no barriers. But it is so expensive that only governments, oil companies, and other giant enterprises can find the necessary capital.

1.3.7 "FEAST OR FAMINE"

The chemical industry is plagued by problems of "feast or famine." The market is cyclical and swings like a pendulum from glut to shortage and back again. The explanation lies in the economies of scale, capital intensity, and ease of market entry that we have already noted. Because of economies of scale a producer will normally install more capacity than he really needs and will rely on exports, expansion of the market, and increase of his share of the home market to sell his surplus. The economies of large scale plants demand, moreover, that they run at close to full capacity most of the time. This will push onto the market more product than is needed at a given price, and accordingly the price will fall. Producers with older, obsolescent plants may find the new price so low that it no longer pays them to produce. When they shut down their plants the supply of product decreases sharply, and users, fearing a shortage, try to build up larger inventories than they would usually carry in order to tide them over. The inventory buildup results in a true shortage in the marketplace. The price rises, and users, fearing further rises, stockpile more product. Eventually user storage capacity is exhausted, and the price has reached a level that deters buyers. Users congratulate themselves on buying when the price was low and start to consume their inventories. Demand for the product drops sharply, and the price declines once more.

Some economists see the need for better planning, perhaps by industry organizations, to alleviate the cycle. This is difficult in the face of laws in the United States that prevent industry collaboration, particularly if there is any hint of price fixing. In other countries there are demands for government intervention.

Other economists, on the other hand, see the cycle as part of the workings of an efficient economy, or at any rate as something like the weather which one cannot and probably would not want to change. They point out that attempts by the UK government to modify the economic cycle over the last 35 years have, if anything, made the cycles worse and that the chemical cartels that existed until World War II cannot, by any stretch of the imagination, be said to have worked in the public interest.

The price fluctuations, furthermore, take place in a narrow range, and examined in the long term prices have been remarkably stable with a tendency to decline. Between 1954 and 1972 the list price of phenol fell steadily from 16 cents lb^{-1} to 7.5 cents lb^{-1}. Ethylene glycol remained steady at about 14 cents lb^{-1} from 1954 to 1968 and at about 9 cents lb^{-1} from 1970 to 1973. Phthalic anhydride, the classic "feast and famine" chemical, on the other hand, shot from 20 cents lb^{-1} in 1958 to 10 cents lb^{-1} in 1966 then up to 13 cents lb^{-1} in 1968, down to 8 cents lb^{-1} in 1971, and up to 19 cents lb^{-1} in 1974. The size of this fluctuation is the exception rather than the rule.

Shortages and inflation together with the fivefold increase in the price of oil created a discontinuity in 1973, and prices doubled or even tripled. There was evidence that a new stability, albeit at a higher price, was being achieved, but the disruption of petroleum supplies caused by the revolution in Iran and the OPEC price increases of June 1979 have thrown chemicals prices into a turmoil again. The measure of price variation is an economic indicator termed "wholesale price index." This is assumed arbitrarily to be 100 for the United States in 1967. It varied between 1950 and 1972 for chemicals and allied products from 88.9 to 104.2. In 1973 the long period of stability started to come to an end, and by 1975 the index was 181.3. By 1976 it had increased to 187.0, as the rate of growth between 1975 and 1976 was slower than for preceding years.

1.3.8 INTERNATIONAL TRADE

Some chemicals can be transported only with danger, difficulty, or at considerable expense. For example, hydrogen, bromine, chlorine, and hydrogen chloride tend to be used at the point of production. Many chemicals, however, can be transported easily and cheaply by truck or in specially designed ships. Ethylene, which for many years was regarded as virtually unshippable, is now available from elaborate pipeline systems linking petrochemical centers in both Western Europe and the United States.

Ease of transport and homogeneity of product mean that there is a large international trade in chemicals. In general the developed countries of the Organization for Economic Cooperation and Development (OECD) trade among themselves, but there are significant exports to developing countries and, of course, huge imports of petroleum from some of them.

Table 1.9 shows chemical exports and imports by developed countries. Most of them have a positive balance of trade. The United States had a positive trade balance of $5 billion which represented a substantial contribution to its balance of payments. On the other hand, exports were only about 8% of the chemical industry's total sales. Other countries rely more heavily on their exports of chemicals. Switzerland exports about 80% of her total production, and the Benelux countries export more than half. The huge West German industry exports more than a third, and the UK industry exports about one-fifth. In 1975 a total of almost $80 billion worth of chemicals were exported by all exporting countries.

Countries outside the OECD, by and large, lack the ability to manufacture sophisticated chemicals to the standards required by OECD coun-

Table 1.9   World Chemical Exports and Imports 1975[a]

	Exports million $	Imports million $	Trade Balance million $
United States	8,705	3,696	+5,009
Canada	1,063	1,772	−709
EEC, nine countries, total	32,203	22,115	+10,088
Belgium-Luxembourg	3,429	2,408	+1,021
France	4,997	4,336	+661
West Germany	10,550	5,280	+5,270
Italy	2,583	2,825	−242
Netherlands	4,998	2,717	+2,281
United Kingdom	4,837	3,129	+1,708
Switzerland	2,759	1,446	+1,313
Japan	3,889	2,057	+1,832

[a] Based on OECD figures (see notes).

tries, and they tend to have a negative trade balance in chemicals. This situation may change in the future as oil-producing countries build their own chemical industries, but it seems likely that OECD countries will stay ahead in the high technology end of the business.

Tariffs or the lack of them can be important to the trade in chemicals and are crucial, for example, in pharmaceuticals. They are also important to the development of chemical industries in the Third World. India is currently forcing the development of its own pharmaceuticals and fine chemicals industries by import restriction policies similar to those employed by the United States in the nineteenth century.

1.3.9 CRITICALITY AND PERVASIVENESS

A chemical industry is critical to the economy of a developed country. In the first half of the twentieth century a nation's industrial development could be gauged from her production of sulfuric acid, the grandfather of economic indicators. Today one uses ethylene production as a yardstick of industrial sophistication. An advanced economy cannot exist without a chemical industry; neither can a chemical industry exist without an advanced economy to support it and to provide the educated manpower it requires.

The chemical industry is not replaceable. There is no other industry that could fulfill its function. It is pervasive and reflected in all goods and services. Not only is the chemical industry here to stay, but also it is a dynamic and innovative industry that has grown rapidly and on which the world will rely in the future. Many of the problems concerning pollution, and energy and raw materials shortages have been detected and monitored by chemical methods, and chemistry will have a part to play in their solutions.

1.4 THE TOP CHEMICAL COMPANIES

Table 1.10 shows the 50 largest chemical and allied products companies in the world excluding the Iron Curtain countries. All but three of the top 100 are based in a mere eight countries (Sec. 1.3.2). Practically all of them also have nonchemical activities. Union Carbide for example is 40% nonchemical.

Petroleum companies are liberally represented in the list. The fifth largest chemical company in the United States, Exxon, is primarily a petroleum company. Its chemical activities comprise only 6% of its total sales, which are nonetheless large enough to give it a distinguished

Table 1.10 The 50 Largest Chemical and Allied Products Companies in the World[a]

Rank 1976	Company	Country	Sales (million $)	Pre-tax Profits (million $)	Research Expenditures (sales %)	Employees
1	Hoechst	W. Germany	9332.9	608.9	4.1	182,980
2	Badische Anilin und Soda Fabrik	W. Germany	9202.6	558.7	2.9	112,686
3	DuPont	US	8361.0	822.1	4.2	132,737
4	Bayer	W. Germany	8297.7	516.2	4.2	171,200
5	ICI[b]	UK	7452.1	973.2	3.3	192,000
6	Union Carbide	US	6345.7	726.5	2.2	113,118
7	Montedison	Italy	6124.7	−248.0	2.4	144,545
8	Dow	US	5652.1	1034.4	3.3	53,000
9	Rhone-Poulenc	France	4520.9	−71.4	4.6	—
10	Monsanto	US	4270.2	617.7	2.6	61,903
11	Akzo	Netherlands	4067.8	21.0	3.8	91,100
12	Veba-Chemie	W. Germany	4034.0	435.1	—	11,427
13	Royal Dutch/Shell	UK/Netherlands	3983.5	—	—	—
14	Ciba-Geigy	Switzerland	3792.4	127.9	7.6	74,355
15	Exxon	US	3741.0	370.0	—	—
16	W. R. Grace	US	3615.1	252.8	0.8	59,700
17	DSM	Netherlands	3521.0	78.6	1.1	32,600
18	Allied Chemical	US	2630.0	189.7	1.6	33,448
19	Warner-Lambert	US	2349.2	295.5	3.4	58,000
20	Solvay	Belgium	2275.3	126.7	1.6	44,701
21	PPG Industries	US	2254.8	278.2	2.5	36,300
22	Celanese	US	2123.0	130.0	3.3	31,900
23	Cyanamid	US	2093.8	205.4	4.0	39,401
24	Hoffmann La Roche	Switzerland	2042.0	189.9	14.8	38,305
25	Bristol-Myers	US	1986.4	282.9	3.5	30,900
26	Mitsubishi Chemical	Japan	1978.8	39.5	—	11,229
27	Pfizer	US	1887.5	263.0	4.7	40,100
28	Sumitomo Chemical	Japan	1834.1	−0.9	2.1	15,697
29	Degussa	W. Germany	1690.0	41.9	2.4	20,000
30	Merck & Co.	US	1661.5	416.4	8.2	27,500
31	Sandoz	Switzerland	1616.5	28.6	9.0	33,400
32	Hercules	US	1595.9	171.0	2.2	23,957
33	Asahi Chemical	Japan	1561.4	14.8	1.5	18,219
34	Occidental	US	1538.0	151.6	1.3	18,600
35	Standard Oil of Indiana	US	1436.0	147.0	—	—
36	Phillips Petroleum	US	1399.7	164.4	—	—
37	Olin	US	1376.5	125.4	1.7	22,000

Table 1.10 (*Continued*)

Rank 1976	Company	Country	Sales (million $)	Pre-tax Profits (million $)	Research Expenditures (sales %)	Employees
38	Diamond Shamrock	US	1356.6	236.3	1.8	10,534
39	Eli Lilly	US	1340.6	341.0	8.4	23,300
40	Snia Viscosa	Italy	1319.1	0.0	1.3	—
41	Beecham	UK	1299.0	228.5	2.1	31,100
42	Int. Minerals & Chemical	US	1260.0	225.6	—	9,674
43	Showa Denko	Japan	1249.9	0.0	—	7,666
44	Eastman Kodak	US	1247.0	—	—	17,800
45	Union Explosives Rio Tinto	Spain	1218.8	41.2	0	20,070
46	Squibb	US	1214.5	155.9	4.1	34,000
47	Teijin	Japan	1152.7	12.5	2.0	12,010
48	Ethyl	US	1135.4	130.9	2.2	16,000
49	Henkel	W. Germany	1112.7	—	—	—
50	Sterling Drug	US	1105.2	153.5	3.3	28,000

[a] Based on *Chemical Age International* (see notes).
[b] Imperial Chemical Industries.

position in the chemical league table. How can this be so in the light of the buildup we have given the chemical industry? There are two reasons. First, the petroleum refining and related products industry is indeed larger than the chemicals and allied products industry (Figure 1.1). Second, the petroleum industry is concentrated in fewer hands than the chemical industry. It is overwhelmingly dominated by the "seven sisters"—seven large vertically integrated companies, Exxon, Mobil, Standard Oil of California, Gulf Oil, Texaco, Royal Dutch Shell, and British Petroleum (BP).

The list includes a number of pharmaceutical companies and general companies such as Imperial Chemical Industries (ICI) or Bayer who have pharmaceutical divisions. Many drug companies have large nonchemical interests. Beecham, for example, is about two-thirds nonchemical, making toilet preparations, drinks and canned foods. Pharmaceuticals are low in volume but high in unit value and this accounts for the appearance in the top 50 of Ciba-Geigy, Warner-Lambert, Hoffmann La Roche, Bristol-Myers, Pfizer, Merck, Sandoz, Eli Lilly, Beecham, Squibb, and Sterling Drug.

Table 1.11   The 50 Highest Volume Chemicals in the United States for 1977[a]

Rank	Chemical	Production (billion lb)	Average Annual Change 1967–1977 (%)
1	Sulfuric acid	68.80	1.9
2	Lime	37.78	1.9
3	Ammonia, anhydrous	32.35	3.3
4	Oxygen, high and low purity	31.86	5.8
5	Ethylene	24.65	10.8
6	Nitrogen, high and low purity	24.04	22.4
7	Chlorine, gas	21.30	3.9
8	Sodium hydroxide	21.00	3.2
9	Sodium carbonate	15.97	4.3
10	Phosphoric acid, total	15.60	5.0
11	Nitric acid	14.77	1.8
12	Ammonium nitrate	13.97	2.2
13	Propylene	12.56	11.8
14	Benzene	11.25	5.8
15	Ethylene dichloride	10.48	16.4
16	Urea, primary solution	8.99	10.6
17	Toluene, all grades	7.73	6.6
18	Ethylbenzene	7.30	11.8
19	Styrene	6.82	10.8
20	Methanol, synthetic	6.46	8.8
21	Formaldehyde, 37% by weight	6.08	6.4
22	Xylene, all grades	6.05	8.4
23	Vinyl chloride	5.81	14.0
24	Hydrochloric acid	5.13	5.8
25	Terephthalic acid	5.01	35.7
26	Carbon dioxide, all forms	4.45	10.5
27	Ethylene oxide	4.42	9.2
28	Ammonium sulfate	3.82	−0.1
29	Carbon black	3.48	4.0
30	Ethylene glycol	3.47	7.4
31	Butadiene (1,3-), rubber grade	3.19	2.0
32	p-Xylene	3.02	29.9
33	Cumene	2.64	13.3
34	Acetic acid, synthetic	2.58	6.5
35	Sodium sulfate	2.51	−0.8
36	Calcium chloride	2.42	0.4
37	Phenol, synthetic	2.38	8.4
38	Aluminum sulfate	2.32	1.2

Table 1.11 (*Continued*)

Rank	Chemical	Production (billion lb)	Average Annual Change 1967–1977 (%)
39	Cyclohexane	2.24	2.6
40	Acetone	2.14	6.7
41	Propylene oxide	1.90	13.3
42	Isopropyl alcohol	1.87	−0.9
43	Adipic acid	1.85	9.1
44	Acrylonitrile	1.64	14.5
45	Vinyl acetate	1.60	16.6
46	Sodium silicate	1.56	2.7
47	Acetic anhydride	1.50	−0.3
48	Titanium dioxide	1.36	1.5
49	Sodium tripolyphosphate	1.30	−3.8
49	Ethanol, synthetic	1.30	−3.2

[a] From *Chemical and Engineering News*, May 1, 1978, p. 33 (see notes).

1.5 THE TOP CHEMICALS

Table 1.11 shows the 50 most important chemicals by volume manufactured in the United States in 1977. The rank order would be more or less the same in any developed country.

Sulfuric acid heads the list by a large margin as befits its position as an economic indicator. Though it has many applications, about 45% is used for phosphate and ammonium sulfate fertilizers. Of the first 12 chemicals, 11 are inorganic. Six of them are associated with the fertilizer industry— sulfuric acid, ammonia, nitrogen, phosphoric acid, nitric acid, and ammonium nitrate. Sodium carbonate is associated with the glass industry; oxygen with the steel industry and welding. This is not to say that these materials are not vital to the organic chemical industry also, but their main markets lie elsewhere.

Ethylene, the most important source of organic chemicals, is the only organic material in the top 12, but the other two major sources of organic chemicals, benzene and propylene, occupy positions 13 and 14. Indeed, of the 38 chemicals that follow the first twelve, 28 are organic, and these medium tonnage chemicals are the backbone of the so-called heavy organic chemical industry. Heavy organics are defined as large volume commodity chemicals such as ethylene and vinyl chloride as opposed to specialty chemicals such as dyes and parmaceuticals.

Some of the chemicals listed have only a single use as a feedstock for another chemical on the list. For example, ethylene dichloride is converted to vinyl chloride, *p*-xylene to terephthalic acid, and cumene to phenol and acetone.

The materials in the list that have shown the highest growth rates over the last decade, with the exception of nitrogen which is used for ammonia production, are all organic chemicals used mainly in the production of polymers. Terephthalic acid and *p*-xylene have grown rapidly because of

**Table 1.12   The 29 Highest Volume Organic
Chemicals Listed According to Sources**[a]

Ethylene	**Benzene**
Ethylene	Benzene
Ethylene dichloride	Ethylbenzene
Ethylbenzene	Styrene
Styrene	Cyclohexane
Vinyl chloride	Phenol
Ethylene oxide	Acetone
Ethylene glycol	Adipic acid
Acetic acid	Cumene
Acetic anhydride	
Ethanol	**Toluene**
Vinyl acetate	Toluene
Propylene	**Xylene**
Propylene	Xylenes (mixed)
Acetone	Terephthalic acid
Isopropanol	*p*-Xylene
Propylene oxide	
Acrylonitrile	**Methane**
Cumene	Urea
Phenol	Methanol
	Formaldehyde
C₄ fraction	Acetic acid
Butadiene	

[a] A chemical that is made from alternative raw materials is listed under both. Thus acetic acid is listed under ethylene and methane. Also a chemical made by combination of two of the above raw materials is listed under both. Thus ethylbenzene is listed under ethylene and benzene.

the boom in polyester fibers. Cumene has grown faster than phenol and acetone because the cumene based route to these materials has been replacing other obsolescent processes.

Propylene oxide has shown high growth because it is the precursor of propylene glycol used in two polymers with high growth rates, unsaturated polyesters and polyurethanes.

The 29 organic chemicals in Table 1.11 are listed again in Table 1.12 according to the feedstock from which they are derived. The total volume in 1976 of ethylene-based chemicals in the list was about 40 billion lb. If we add to this the volume of other ethylene-based chemicals, the ethylene-based polymers, and ethylene itself we have an estimated total of 110 billion lb, or about 44% of all organic chemicals produced. Chemicals based on propylene amount to about 11.8 billion lb. With propylene itself, propylene-based polymers, and other chemicals based on propylene the total is an estimated 45–50 billion lb. About 23 billion lb of chemicals are based on benzene. If to this figure is added the volume of benzene production, benzene-based polymers, and other benzene chemicals the total is an estimated volume of 50–55 billion lb. About 20 billion lb of chemicals are based on methane. Addition of all these figures plus the volume of chemicals based on and including toluene (8 billion lb), xylene (16 billion lb), and butadiene (3 billion lb) provides a total very close to the 250 billion lb of chemicals and polymers the industry is estimated to produce yearly.

There is considerable double counting in the figures in that feedstocks, intermediates, and final products are all added together. Also chemicals like ethylbenzene have been counted under both ethylene and benzene. Nonetheless the statistics serve to indicate that the organic chemical industry depends overwhelmingly on the seven raw materials listed in Table 1.12. The industry has developed an impressively extensive and sophisticated chemistry to enable it to build a huge business on the basis of these few feedstocks.

REFERENCES AND NOTES

The position of the chemical industry in a developed economy is discussed in *The Chemical Economy* and in a slim paperback by S. Hays, *The Chemicals and Allied Industries*, Heinemann, London, 1973. General information about the chemical industry—including sales, volumes, pollution and environmental problems, and trends—can be found in *MCA News* published by the Manufacturing Chemists Association, 1825 Connecticut Ave., Washington, DC 20009. Information is also published annually in the more available *Fortune* and *Chemical and Engineering News*. A statistical analysis of the US chemical industry is contained in "Kline Guide to the Chemical Industry," *Kline Industrial Marketing Guide*, IMG-13-77, 3rd ed., Charles H. Kline & Co., Inc., Fairfield, NJ 07006. The guide is updated at intervals.

Sec. 1.1   Table 1.1 is based on the US Bureau of the Census, Annual Survey of Manufactures, 1975. Use of drugs to render the mentally well insane is described in S. Bloch and P. Reddaway, *Russia's Political Hospitals—The Abuse of Psychiatry in the Soviet Union*, Gollancz, London, 1977.

Sec. 1.2   Figures 1.1 and 1.2 are based on US Bureau of the Census figures and the UK Census of Production. UK figures are modified to make them more comparable with US data.

Sec. 1.3.1 The effect on research and development expenditures of government regulations has been described for General Motors in an article by P. F. Chenea, *Res. Manage.*, March, 22 (1977).

Sec. 1.3.2 Tables 1.6 and 1.7 were compiled from published statistics. Table 1.8 is based on *Fortune's* Top 500 for 1977. About 65% of all chemists and chemical engineers in industry are employed by traditional chemical companies. The remaining 35% are employed by companies who are not in the chemical industry but require the services of chemists. Polaroid, 3M, Bell Telephone, and IBM are examples.
            Details of the world's top 200 chemical companies are published each year in *Chemical Age International* as a supplement to the final issue in June.
            "Technology push" inventions are quite rare and tend to be solutions in search of a problem. See J. Schmookler, *Invention and Economic Growth*, Harvard University Press, Boston, 1966.

Sec. 1.3.4 Economies of scale are discussed in more detail in B. G. Reuben, "Economies of Scale or Diminishing Returns," *Process Eng.*, November, 1974, p. 100. Table 1.8 is based on *Fortune Top 500 for 1977*.

Sec. 1.3.7 Graphs of US prices of individual chemicals over approximately a 20-year period are to be found in Faith, Keyes, and Clark, *Industrial Chemicals*, 4th ed., Wiley Interscience New York, 1975. The US wholesale price indexes for chemicals (now called producer price indexes) appear in *Chem. Eng. News*, **56,** No. 24 (1978), 52, while the UK figures appear in the *Annual Abstract of Statistics*, HMSO, London.

Sec. 1.3.8 Table 1.9 is based on *the Chemical Industry 1975*, OECD, Paris, 1977.

Sec. 1.4   Table 1.10 comes from the *Chemical Age International*, June 24, 1977, supplement.

Sec. 1.5   Table 1.11 comes from *Chem. Eng. News*, **56,** (1978), 33.

Chapter Two
CHEMICALS FROM NATURAL GAS AND PETROLEUM

Where do industrial organic chemicals come from? Table 2.1 provides a guide. Natural gas and petroleum are the main sources. From them come seven chemical building blocks on which a vast organic chemical industry is based. These are ethylene, propylene, the butylenes, benzene, toluene, the xylenes, and methane. The olefins—ethylene, propylene, and the butylenes—are derived from both natural gas and petroleum. The aromatics, benzene, toluene, and xylene, are derived from petroleum and to a much smaller extent from coal. Methane comes from natural gas.

Whether natural gas or petroleum is used for olefins varies throughout the world depending on the availability of natural gas and the demand for gasoline. Both light and heavy naphthas are petroleum fractions that can be cracked to make olefins. They can also be used for gasoline. In the United States demand for gasoline is higher than for other petroleum fractions, and the price of naphtha is therefore high. Consequently the chemical industry has chosen to extract ethane and the higher alkanes for cracking to olefins from what has hitherto been abundant natural gas.

In Western Europe the demand for gasoline is lower because on a per capita basis there are half as many cars as in the United States, and each of these uses half as much gasoline as its American counterpart. Thus in

Table 2.1 Sources of Organic Chemicals

Natural gas and petroleum
Coal
    Acetylene
    Water gas
Fats and oils
    Fatty acids
    Fatty nitrogens
Carbohydrates
    Cellulose
    Fermentation

Western Europe far more naphtha is produced than is required for gasoline. Also, natural gas is less abundant and contains less ethane. Consequently light naphtha has become the important European raw material for the manufacture of olefins. This is equally true in Japan which lacks resources of either natural gas or oil.

Probably 90% by weight of the organic chemicals the world uses come from petroleum and natural gas, and we therefore devote considerable space to them. In addition, we consider what might happen if natural gas and petroleum supplies are exhausted in the next 30–60 years. After all, it has been authoritatively predicted that the United States will never again bring to the surface as much natural gas as she did in 1973. The need for strategies for the future looms large.

A less important source of chemicals is coal. Coal was historically very important, and much of the progress in the chemical industry until World War II was motivated by the availability of coal. Indeed the famous English chemist, W. H. Perkin, could claim to have founded the organic chemical industry in 1856 when he synthesized a dye with a mauve color from coal tar intermediates. So important was Perkin's dye that its color gave its name to a period of history known in the literature as the "mauve" decade.

The decline of coal coincided with the rise of petrochemicals. Reserves of coal are much greater than those of oil. If petroleum becomes scarce, will coal come into its own again? Is it a realistic part of an alternative strategy?

The third and final source of organic chemicals is the group of naturally occurring, *renewable* materials of which fats, oils, and carbohydrates are the most important. Although they account at present for only a few percent by weight of the products of the chemical industry, there has been much discussion of how they might be used to replace nonrenewable fossil materials (oil, coal, natural gas) if the latter ran out. We discuss this too.

The third group also includes more obscure natural products that make contributions to highly specialized segments of the chemical industry. Examples of such materials are sterols, alkaloids, phosphatides, rosin, shellac, and "gums" like gum arabic. Their, contribution to the chemical industry in terms of value or weight is small, and we do not say much about them.

2.1 PETROLEUM DISTILLATION

To gain an idea of how petroleum is used as a source of chemicals we must consider what happens in a petroleum refinery. Let us consider a simple refinery (Fig. 2.1) in which crude oil, a sticky, viscous liquid with an unpleasant odor, is separated by distillation into various fractions.

The first most volatile fraction consists of methane and higher alkanes and is similar to natural gas. The methane can be separated from the higher alkanes, primarily propane and the butanes. The latter, called liquified petroleum gas (LPG), may be used as a petrochemical feedstock (Sec. 2.3) or a fuel.

In the past the high cost of compressing or liquefying and shipping these refinery gases has dictated that most of them be flared. As the price of natural gas increases, however, shipping of methane in refrigerated tankers may become commonplace. Alternatively, methane may be converted to methanol, which is a useful organic chemical (Sec. 2.12) and more easily shipped. Even today, however, most refinery gas is still flared even in the United States.

The second and third fractions are called light naphtha, or straight run gasoline, and heavy naphtha, respectively, and are of particular importance to the chemical industry. The term naphtha is not well-defined, but the material cracked for chemicals generally distills in a range between 30 and 200°C and contains C_5–C_{10} hydrocarbons. Naphtha contains aliphatics as well as cycloaliphatic materials such as cyclohexane, methylcyclohexane, and dimethylcyclohexane. Smaller amounts of polymethylated cycloalkanes and polynuclear compounds such as methyldecahydronaphthalene are also present. Like the lower alkanes naphtha may be cracked to low molecular weight olefins, and this is done in Europe, particularly if light naphtha is in short supply. Its conversion by a process known as catalytic reforming into benzene, toluene, and xylenes (BTX) is very important. In the United States this is its main chemical use. Catalytic reforming is also the major source of aromatics worldwide.

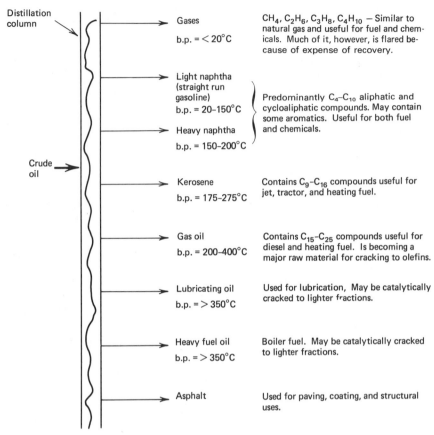

Figure 2.1 The distillation of crude oil.

Catalytic reforming has become even more important in the United States because of the ban on leaded gasoline for new cars. Lead tetraethyl is normally added to gasoline as an octane improver. In its absence octane ratings can be maintained by increasing the proportion of aromatics in the fuel, and demand for aromatics has consequently increased (see notes).

Light naphtha was at one time used directly as gasoline; hence its alternate name, "straight run gasoline." It contains a large proportion of straight chain hydrocarbons (n-alkanes), and these resist oxidation much more than branched chain hydrocarbons (isoalkanes), which contain tertiary carbon atoms. Consequently, straight run gasoline has poor ignition characteristics and a relatively low octane number. It is of little use in modern high compression-ratio automobile engines. Chemically, however, its significance is like that of heavy naphtha, for it can be cracked and reformed to low molecular weight olefins and to BTX.

The situation is summarized in Figure 2.2. International practice differs, however. In the United States the high demand for gasoline means that very little of the available naphtha is used for cracking to olefins. Under these conditions it is more economical to make olefins from the ethane and propane in natural gas. In Western Europe and Japan, however, there is more than enough naphtha to satisfy the gasoline market*. The remainder then is thermally cracked (Sec. 2.2.1) to olefins, reformed to BTX, and a small amount is even oxidized directly to acetic acid (Sec. 2.4.3).

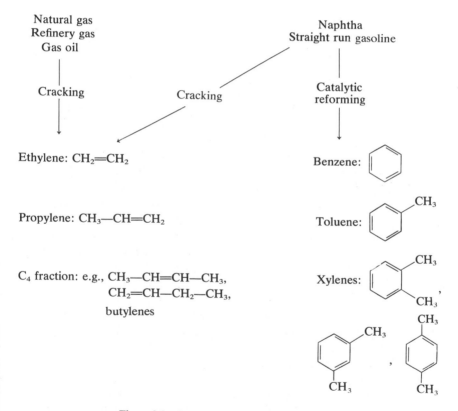

Figure 2.2 Cracking and catalytic reforming.

* In recent years, the chemical industry has been making increasing demands on naphtha supplies. By 1979 there was a worldwide shortage of naphtha which was part of a general shortage of petroleum products.

Table 2.2 Sources of US and European Ethylene[a]

| | United States (%) | | |
	1970	1976	Europe (%)
Ethane ⎫		53 ⎫	
Propane ⎬	75	19 ⎬	6
Butane ⎭		5 ⎭	
Naphtha ⎱	10	8	90
Gas oil ⎰		11	Small
Refinery gas	15	4	4

[a] See notes.

The point having been made, it must now be qualified. Shortage of natural gas in the United States has led to an increased interest in liquid feedstock cracking. However, in the United States it is not naphtha that will be cracked, except in unusual circumstances, but rather the next higher fraction, gas oil. Gas oil cracking was not highly regarded when natural gas was plentiful for two reasons. First, it could be catalytically cracked to gasoline fractions. Second, thermal or steam cracking to olefins was accompanied by tar and coke formation. This latter problem, under the creative impetus of necessity, has apparently been moderated. By 1978, (1970–1976 figures in Table 2.2) it is estimated that liquid feedstocks will account for 35–40% of all feedstock for ethylene in the United States, whereas natural gas sources will have decreased to 50–55%. The amount of liquid feedstock is predicted to rise to 60% by 1990. In addition, the European picture may change. The Brent field oil in the northern region of the North Sea appears to have significant amounts of associated gas containing C_2–C_5 compounds. This contrasts with the early gas strikes off the Lincolnshire Coast which were virtually pure methane. An ethane separating plant is being built, and a cracker is planned for the early 1980s, which may form the basis for a Scottish petrochemical industry. Obviously the US and European patterns are coming closer together. The burgeoning petrochemical industry in the Mideast plans to produce olefins from naphtha. Furthermore, a way is being developed to crack whole petroleum to olefins, and in the long run this may prove economical.

Two reactions in Figure 2.2—steam cracking and catalytic reforming—account for the sources of much of the world's petrochemical production valued in 1974 at \$106 billion. The three main raw materials are ethylene, propylene, and benzene while the C_4 fraction, methane, toluene, and the xylenes are important but to a lesser degree. Methane is an important source not only of organic chemicals but also of ammonia (Sec. 2.12).

Returning to petroleum refining (Fig. 2.1) we can see that the naphtha fractions occupy the uncomfortable position of being a source of both fuel and chemicals. Kerosene is a fuel for tractors, jet aircraft, and for domestic heating and has some applications as a solvent. Gas oil is further refined into diesel fuel and light fuel oil of low viscosity for domestic use. Its new and important application in the United States is as feed for cracking units for olefin production. Both the kerosene and gas oil fractions may be catalytically cracked to gasoline range materials.

The residual oil boiling above 350°C contains the less volatile hydrocarbons together with asphalts and other tars. Most of this is sold cheaply as a high viscosity heavy fuel oil (bunker oil) which must be burned with the aid of special atomizers. It is used chiefly on ships and in industrial furnaces.

A proportion of the residual oil is vacuum distilled at 0.07 bar to give fuel oil (b.p. < 350°C), wax distillate (350–560°C), and cylinder stock (> 560°C). The cylinder stock is separated into asphalts and a hydrocarbon oil by solvent extraction with liquid propane in which asphalts are insoluble. The oil is blended with the wax distillate, and the blend is mixed with toluene and methyl ethyl ketone and cooled to −5°C to precipitate "slack wax," which is filtered off. The dewaxed oils are purified by countercurrent extraction with such solvents as furfural, which remove heavy aromatics and other undesirable constitutents. The oils are then decolorized with Fuller's Earth or bauxite and are blended to give lubricants (Part II, Chapter 10).

Part of the vacuum distillate and the "slack wax" can be further purified to give paraffin and microcrystalline waxes used for candles and the impregnation of paper. The petroleum industry is constantly trying to find methods by which the less valuable higher fractions from petroleum distillation can be turned into gasoline or petrochemicals.

2.2 PETROLEUM REFINING REACTIONS

The production of chemical feedstocks from petroleum is inextricably associated with the production of gasoline and other fuels. Sometimes these two industries compete for raw materials, and sometimes they complement one another. The chemical industry is a junior partner because in few countries does it consume as much as 10% of refinery output. On the other hand, because it produces premium products, it can compete with other consumers in buying those raw materials it needs. The reason why the chemical industry can compete successfully is illus-

---

CRACKING: A A process for making molecules with 5–12 carbon atoms from larger molecules.

             B A process for converting saturates like ethane or propane into ethylene or propylene.

Thermal Cracking: Used for B above. The modern version of the process is also called steam cracking.

Catalytic Cracking: Used for A above. Facilitates formation of branched chain and aromatic molecules.

Hydrocracking: Uses catalysts together with hydrogen which prevents "coking" of catalyst and converts objectionable sulfur, nitrogen, and oxygen compounds to volatile H_2S, NH_3, and H_2O.

POLYMERIZATION: Combines low molecular weight olefins into gasoline-range molecules with H_2SO_4 or H_3PO_4 catalysts. Not widely used today.

ALKYLATION: Combines an olefin with a paraffin (e.g., propylene with isobutane) to give branched chain molecules. H_2SO_4 and HF are used as catalysts. Very important in achieving high octane number in lead-free gasoline.

CATALYTIC REFORMING: Dehydrogenates both straight-chain and cyclic aliphatics to aromatics, primarily benzene, toluene, and xylenes (BTX) over platinum-rhenium-alumina catalysts. Most widely used refinery reaction. Very important to achieve high octane number in lead-free gasoline.

DEHYDROGENATION: Cracking and reforming are basically dehydrogenations. Other dehydrogenations include conversion of ethylbenzene to styrene, and butane to butadiene.

ISOMERIZATION: Used to convert straight chain to branched chain compounds—for example, *n*-butane to isobutane for alkylation. Other isomerizations include ethylbenzene, *o*-xylene, and *m*-xylene to *p*-xylene.

COKING: Used to remove metals from a refinery stream. Heat in the absence of air "cracks-off" hydrocarbons. The metals stay behind in the "coke."

HYDROTREATING: Used to convert sulfur, nitrogen, and oxygen in petroleum fractions to the volatile gases, H_2S, NH_3, H_2O. Uses hydrogen from other refining processes such as reforming. Can be applied to heavy feedstocks.

Figure 2.3 Petroleum refining reactions.

trated by the following tabulation*:

Form of Oil	Value of oil
As crude oil	X
As fuel	2X
As a petrochemical (average)	13X
As a consumer product (average)	55X

In order to understand this competition and the operation of a petroleum refinery we must examine processes other than distillation.

We have already noted that straight run gasoline has too low an octane number for high compression-ratio engines. A major objective of a petroleum refinery is to raise this number. It is achieved by way of the reactions summarized in Figure 2.3 which either modify a petroleum fraction or provide the raw material for another reaction that will give compounds with an improved octane number.

In a modern refinery these reactions take place under computer control, and their proportions are varied according to the ever-changing demands of the market and the composition of the feedstock, which may vary from day to day and storage tank to storage tank.

2.2.1 CRACKING AND REFORMING

Two of the reactions already mentioned, catalytic cracking and catalytic reforming, take place under the influence of catalysts. Synthetic or natural silica-aluminas are used for cracking, whereas platinum or platinum-rhenium on alumina are used for reforming.

Catalytic cracking, as its name implies, fragments molecules, and it also dehydrogenates some of them. Temperatures of about 450–550°C are used. Typical heavy naphtha and gas oil feedstocks give a mixture of products consisting mainly of isoalkanes suitable for gasoline together with n- and iso-olefins and n-alkanes from C_3 upwards. Little ethylene is produced although significant amounts of propylene form (Sec. 2.5). Though important to the gasoline industry, catalytic cracking is not a major route to petrochemicals.

Catalytic reforming leaves the number of carbon atoms in the feedstock molecules unchanged; for example, cyclohexane goes to benzene. The process however is more complex than that. Not only may substituted cyclohexanes be converted to substituted benzenes and straight chain

*Taken from *Chem. Eng. News*, March 29, 1976, p. 6.

paraffins such as n-heptane cyclized to aromatics such as toluene, but also substituted cyclopentanes may be ring-expanded to aromatics (Sec. 2.6). If heavy naphthas are used as feedstock, methylnaphthalenes are formed. Like catalytic cracking, catalytic reforming is a carbonium ion reaction, but the reactions leading to aromatics are favored.

Catalytic reforming of the C_6–C_8 fraction of light naphtha at 400°–500°C and 25–35 bar gives a gas yield of about 15% by weight consisting of hydrogen and the lower alkanes. The liquid products consist of a high octane gasoline, rich in aromatics and low in olefinic compounds which would give gums in contact with air. It may be used to upgrade other gasoline or as a source of BTX, which makes up about half the products. The remainder is C_5 alkanes and aromatics above C_8.

The separation is complicated. The gas stream from the reformer is cooled, and products containing five or more carbon atoms condense. Hydrogen and the C_1–C_4 alkanes are taken as a top product. The hydrogen is usually used to dealkylate toluene (Sec. 2.8), and the alkanes are burned as fuel.

The liquid product is treated with a solvent that preferentially dissolves aromatic compounds. Diethylene glycol/water, N-methylpyrrolidone/ethylene glycol, and sulfolane are used in various processes. The aromatics appear in the extract, and the C_5 compounds are left in the raffinate. The solvent is distilled off and recycled, and the benzene, toluene, and mixed xylenes are separated on three fractional distillation columns, leaving a high boiling C_9+ aromatics fraction. Separation of the mixed xylenes is difficult because their boiling points are so close. Orthoxylene (b.p. 144.4°C) is separated from m-xylene (b.p. 139.1°C) and p-xylene (b.p. 138.3°C) on a very high ($\sim$200 plates) distillation column. The m- and p- isomers are then cooled to −60°C. Pure p-xylene (m.p. 13.2°C) crystallizes out, leaving a liqor rich in m-xylene (m.p. −47.9°C).

Variation of reaction conditions and catalyst and introduction of further isomerization steps allows a degree of choice in reaction products. For example, p-xylene has traditionally been more valuable than its isomers, m-xylene, o-xylene, and ethylbenzene, because it can be oxidized to terephthalic acid, which is the basis for "Dacron" or "Terylene" polyester fibers. Reforming and isomerization processes have therefore been developed so that the C_8 fraction from which the p-xylene has been removed can be converted into an equilibrium mixture containing more of the p-isomer. After its removal the residue can again be isomerized and the process continued indefinitely.

Hydrocracking is a variant of catalytic cracking in which a different catalyst is used, and the cracking reactions take place in an environment of hydrogen at 60–100 atm pressure. Because hydrogen is present the catalyst does not "coke" as it does in catalytic cracking, a wider range of feedstocks can be tolerated (e.g., heavy distillates can be used), and objectionable sulfur, nitrogen, and oxygen compounds are converted to hydrogen sulfide, ammonia, and water. The products are paraffins, not olefins, and are fairly low in aromatics and very low in sulfur. They are used for low-sulfur jet fuels and diesel fuels where lack of aromatics and absence of sulfur are desirable.

Thermal cracking, introduced as early as 1912, involves high temperatures of approximately 800°C, but no catalyst. From naphtha feedstocks it gives yields of C_2, C_3, and C_4 olefins and relatively low yields of gasoline. Consequently it has long since been superseded for gasoline production by catalytic cracking and was only revived when the production of cheap ethylene became the central concern of the infant petrochemical industry. It is now widely used on both naphtha and alkane feedstocks, and its use on gas oil is increasing.

There is little similarity between the old and new thermal cracking processes. The obsolete process used a heavy feedstock and a relatively low temperature and high pressure to maximize gasoline production and minimize gas formation. The new process uses light liquid or gaseous feedstocks, a high temperature (850–900°C), and a low pressure to maximize the yield of gases. Since it is inconvenient to operate a plant below atmospheric pressure because a small leak could lead to the formation of explosive hydrocarbon/air mixtures, the partial pressure of reactants is reduced by addition of steam as an inert diluent. The steam also serves by way of the water-gas reaction (Sec. 2.12) to reduce coking of the catalyst. Coking is potentially a major problem and is also reduced by use of short residence times. In the United States ethane and propane from natural gas are thermally cracked to ethylene and propylene. The insignificance of chemicals production compared with that of gasoline is illustrated by the fact that only 9% of total US refinery cracking capacity is thermal cracking, yet this is the only way in which ethane and propane from natural gas can be converted to olefins. Cracking of gas oil also is thermal.

Not surprisingly, the products of thermal cracking are affected by feedstock as well as by reaction conditions. Modern crackers operate under severe conditions (high temperatures, low residence times), and in the last quarter century the typical ethylene yield by weight from fairly severe cracking of a naphtha feedstock has been approximately doubled from about 16 to 33%. As ethylene is the premium product, this is a major advance.

The effect of feedstock is illustrated in Table 2.3. The economics of the different feedstocks depend crucially on whether products other than

Table 2.3   Yields by Weight from Cracking of Various Feedstocks (Including Ethane Recycle)[a]

Feed	Products (wt %)				
	Ethylene	Propylene	Butadiene	BTX	Other[b]
Ethane	84	1.4	1.4	0.4	12.8
Propane	44	15.6	3.4	2.8	34.2
Naphtha (severe cracking)	31.7	13.0	4.7	13.7	36.9
Naphtha (mild cracking)	25.8	16.0	4.5	10	43.7
Gas oil	28.3	13.5	4.8	10.9	42.5
Crude oil[c]	32.8	4.4	3.0	14.4	45.4

[a] Derived from L. F. Hatch and S. Mattarin *Hydrocarbon Proc.* (see notes).
[b] Other products include methane, butylenes, C_5 and above aliphatic hydrocarbons, and fuel oil.
[c] Process still in the pilot plant stage.

ethylene can be credited at premium prices or have to be credited at fuel value. That in turn depends on whether the cracker is operated on a large enough scale for it to be worthwhile to separate individual by-products and whether the cracker is part of a large integrated petrochemical complex that can use all the side streams it produces.

It would normally be possible to illustrate the economics of cracking and reforming by a price comparison. The petroleum industry price structure, however, underwent major upheavals in winter 1973 and spring 1979 and prices have fluctuated wildly. Our comparisons, therefore, must be largely qualitative.

1   Crude oil and refinery products are cheaper in the United States than in Europe or Japan. In June 1979, Saudi Arabian light 34° crude was selling at $132 tonne^{-1}, while United States crude was price-controlled at $43 tonne^{-1} to certain customers and $96 tonne^{-1} to others. Within this price structure, there are relative differences, naphtha being relatively cheap in Europe and Japan.

2   In the United States, naphtha is normally dearer than gas oil. Ethane and propane are not refinery products and their prices are not linked to the petroleum price; in 1978 the naphtha price overtook the ethane and propane prices. In Europe gas oil became dearer than naphtha in 1978 and ethane and propane were not available.

3 Ethylene, butadiene, benzene and p-xylene are the premium products. Propylene and the butenes are cheaper and 2-butene is cheaper than 1-butene. Mixed xylenes have a price similar to toluene, which is cheaper than o-xylene, which in turn is cheaper than p-xylene.

The costs of operation of a cracker also depend on the complexity of the separation processes involved. Ethane crackers produce so few by-products that separation is simple; naphtha and gas oil crackers are more complicated and consequently more expensive.

Conventionally, the product gases from a cracker are distilled at atmospheric pressure, and fuel oil and cracked gasoline (the C_5+ products) are removed as residue. The gases are then scrubbed with sodium hydroxide to remove acid gases such as carbon dioxide and hydrogen sulfide, and dried with molecular sieve. The remaining gases are compressed to 40 bar and distilled in a column called a demethanizer, which carries a condenser cooled to $-95°C$ with liquid ethylene. Methane and hydrogen are the top products, and the C_2+ products emerge as a liquid at the bottom. This bottom stream is distilled again in a deethanizer in which the C_2 components pass overhead and the $C_3–C_5$ hydrocarbons form the bottom product. The C_2 fraction is selectively hydrogenated to reduce traces of acetylene to ethylene, and the ethylene/ethane mixture is distilled to give an ethylene top product that may be further purified to give polymer grade ethylene, and an ethane bottom product that is recycled. A depropanizer column separates the C_3 and C_4 streams, and a debutanizer separates the C_4's from C_5's. The latter are amalgamated with the cracked gasoline stream from the first column. Butadiene and isoprene are recovered from the C_4 and C_5 streams by extractive distillation with acetonitrile.

The size of these various facilities depends on the proportions of the various products. In spite of elaborate systems of heat exchangers to economize on energy and the use of centrifugal rather than reciprocating compressors, the separation processes are costly to build and operate. Ethane gives the fewest by-products and is clearly the preferred feedstock if available at a reasonable cost.

2.2.2 MECHANISM OF CRACKING

Catalytic cracking involves carbonium ion intermediates, but thermal cracking, which is of greater concern to the chemical industry, is a free radical reaction involving the steps illustrated in Figure 2.4. n-Decane is a molecule that occurs in gas oil. If a carbon-carbon bond breaks, two

a $CH_3-CH_2-CH_2-CH_2-CH_2-CH_2-CH_2-CH_2-CH_2-CH_3 \xrightarrow{heat}$

n-Decane

$$2CH_3-CH_2-CH_2-CH_2-CH_2\cdot$$

b (1) $CH_3-CH_2-CH_2-CH_2-CH_2\cdot \longrightarrow CH_3-CH_2-CH_2\cdot + CH_2{=}CH_2$

Ethylene

(2) $CH_3-CH_2-CH_2\cdot \longrightarrow CH_3\cdot + CH_2{=}CH_2$

c $CH_3\cdot + RH \longrightarrow CH_4 + R\cdot$

d $CH_3-CH_2-CH_2-CH_2-CH_2-CH_2-\overset{\cdot}{C}H-CH_2-CH_3 \longrightarrow$

$$CH_3-CH_2-CH_2-CH_2-CH_2\cdot + CH_2{=}CH-CH_2-CH_3$$

1-Butene

e $CH_3-CH_2\cdot \longrightarrow CH_2{=}CH_2 + H\cdot$

f $\text{\small\textasciitilde\textasciitilde\textasciitilde}CH_2\cdot + \cdot CH_2\text{\small\textasciitilde\textasciitilde\textasciitilde} \longrightarrow \text{\small\textasciitilde\textasciitilde\textasciitilde}CH_2-CH_2\text{\small\textasciitilde\textasciitilde\textasciitilde}$

Figure 2.4 Cracking reactions.

radicals form (**a**). A free radical chain reaction may start and [**b**(1) and **b**(2)] the radicals undergo β-scission to yield new free radicals and ethylene. Equation **c** shows a hydrogen abstraction from a methyl radical to create a new free radical. It is this new free radical that gives rise to olefins other than ethylene. The alkyl radical of **c** is practically always a secondary radical as might be expected both statistically and from energy considerations. When this radical (**d**) undergoes β-scission molecules other than ethylene, for example 1-butene, will be produced. The other product, the free radical, can react as in **b**. Carbon-hydrogen β-scission may also occur to yield ethylene (**e**), and this accounts also for the hydrogen released during cracking.

Equation **f** shows a termination reaction by radical coupling. Both this reaction and reactions **c** and **d** prevent high yields of ethylene. The reactions **c** and **f** are bimolecular and are inhibited by low partial pressure. This is one reason why steam is included. Higher temperatures favor both the chain initiation reaction **a** and the β-scission reactions **b**(1) and **b**(2). Equation **f**, which has a zero activation energy, is temperature independent. Thus high temperatures favor formation of primary radicals and β-scission reactions, which in turn contribute to higher ethylene yields.

2.2.3 POLYMERIZATION

The next reaction in the summary listed in Figure 2.3 is polymerization. Low molecular weight olefins produced by cracking are dimerized and trimerized into species that can be used as gasoline. This was important for the production of aviation fuels during World War II and has more recently been used for production of propylene trimer and tetramer (Part II Sec. 7.6) for conversion to alkylbenzenes, which in turn were converted to detergents by sulfonation.

2.2.4 ALKYLATION

Polymerization, which involves two moles of olefins, has been replaced by the more sophisticated alkylation reaction between one mole each of a paraffin and an olefin. An olefin such as isobutene is used to alkylate a branched chain hydrocarbon such as isobutane in the presence of Friedel-Crafts catalysts such as sulfuric acid or HF:

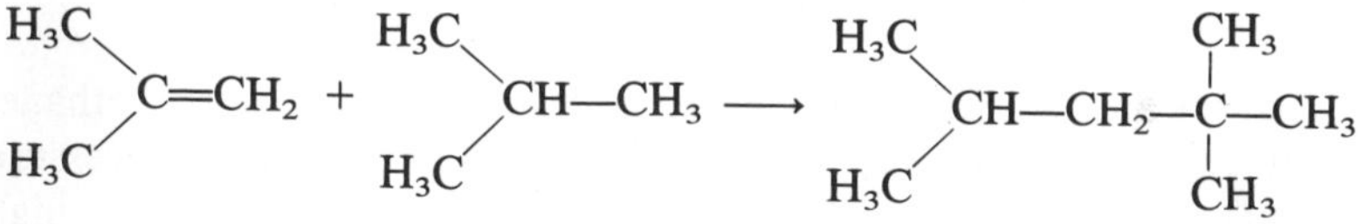

$$
\underset{H_3C}{\overset{H_3C}{>}}C{=}CH_2 \;+\; \underset{H_3C}{\overset{H_3C}{>}}CH{-}CH_3 \;\longrightarrow\; \underset{H_3C}{\overset{H_3C}{>}}CH{-}CH_2{-}\underset{CH_3}{\overset{CH_3}{\underset{|}{\overset{|}{C}}}}{-}CH_3
$$

2,2,4-Trimethylpentane ("isooctane")

2,3,4- and 2,3,3-trimethylpentanes are also formed. The interaction of propylene with isobutane yields a complex mixture including 38% 2,3-dimethylpentane and 19% 2,4-dimethylpentane. In addition, 25% propane results. The remainder is a mixture of branched chain hydrocarbons with eight carbon atoms. Since the carbonium ion intermediate attacks at tertiary carbon atoms, the products are highly branched and have high octane numbers. The octane number of a fuel is defined in terms of its knocking characteristics relative to n-heptane and isooctane, which have been arbitrarily assigned octane numbers of 0 and 100 respectively. The octane number of an unknown fuel is the volume percent of isooctane in a blend with n-heptane that has the same knocking characteristics as the unknown fuel in a standard engine. Alkylation is extremely important in a modern US refinery.

2.2.5 DEHYDROGENATION, ISOMERIZATION, COKING, AND HYDROTREATING

The term dehydrogenation may be used to describe all the forms of cracking and reforming. In refinery practice, however, its use is restricted to specialized dehydrogenations such as conversion of ethylbenzene to styrene, or butenes to butadiene. If cracking and reforming are considered as dehydrogenations, then dehydrogenation is one of the most important, if not the most important, reaction carried out industrially.

Isomerization is used to convert straight to branched chain compounds. The term may be applied to cracking and reforming reactions but usually these are excluded. In petroleum refining n-butane is isomerized to isobutane to supplement other supplies of isobutane for alkylation. Isomerization is also used to increase the octane rating of C_5/C_6 alkanes and to convert a mixture of m- and o-xylenes and ethylbenzene from which the p-xylene has been extracted into a random mixture of all four compounds from which the p-xylene can again be extracted (Sec. 2.9).

The last two reactions in Figure 2.3, coking and hydrotreating, are methods devised to remove contaminants—metals, sulfur, nitrogen, and oxygen—from feedstocks. These may interfere with the refining reactions, poison the catalysts, and give rise to pollution.

2.3 SEPARATION OF NATURAL GAS

Most American natural gases contain recoverable amounts of ethane, propane, and higher alkanes. At the wellhead the gas is at a high pressure (30–100 atm), and propane and higher alkanes may be absorbed in a high boiling oil at ambient temperature and subsequently purified by low temperature fractional distillation. Ethane may be absorbed similarly at −50°C. Hydrogen sulfide and carbon dioxide are scrubbed out with aqueous mono- or diethanolamine and water removed with hygroscopic diethylene glycol.

The organic chemical industry thus competes, although on a fairly small scale, for a number of products obtained from natural gas extraction and petroleum refining that would otherwise be used as fuel. These are listed in Table 2.4.

Methane is the major component of natural gas and may also be "steam-reformed" (Sec. 2.12) to synthesis gas. Ethane is also a component of natural gas, especially in the United States where heating gas, by law, must contain sufficient ethane to maintain the Btu content at a specified level. If these regulations were relaxed a large amount of ethane would become available for ethylene production, and Btu content would be affected only slightly.

Similarly, liquid petroleum gas consists of propane and butane that could otherwise be cracked to propylene and butylene. These two olefins may in turn be used for alkylation reactions with isobutane rather than as chemical raw materials. Finally, the BTX fraction is required not only for chemicals but also for raising the octane number of gasoline.

Table 2.4 Chemicals and Fuel Compete for Raw Materials

	Fuel	Chemicals
Methane	Major component of natural gas	Source of synthesis gas, $CO+H_2$; ammonia
Ethane	Raises the Btu content of heating gas	Source of ethylene, propylene, and butylenes
Propane and butane	Basis for liquid petroleum gas, (LPG)	
Propylene	Used in alkylation reactions to prepare high octane number, branched chain hydrocarbons	Starting materials for chemicals and polymers
Butylene	As for propylene	
Benzene, toluene, and xylenes	Used to increase octane number of gasoline	

In the United States, therefore, the chemical industry has relied traditionally on natural gas producers for feedstocks for olefin production, and by using much of the ethane and propane and a little of the methane, consumes about 10% of their total volume. It relies on refineries for some gases such as propylene that are by-products of catalytic cracking and for aromatics which, of course, are also required by gasoline users. About 3% of US petroleum feedstocks are used for chemicals. On the average, 6–6.5% of total natural gas and petroleum is converted to chemicals.

In Western Europe and Japan the chemical industry relies on natural gas for methane but at present for nothing else. Olefins and aromatics are all obtained by thermal cracking and catalytic reforming of light naphtha. The main demand in these countries is for fuel oil, and the chemical industry and gasoline users are allies in helping to consume the gasoline surplus.

These differences are important in optimum allocation of resources, and many contrasts between the United States and the rest of the developed world can be understood in terms of them.

2.4 CHEMICALS FROM ETHYLENE

We have described how the petroleum refinery and the light hydrocarbon cracking units provide the seven raw materials on which the petrochemical industry is based. We can now examine the extensive chemistry associated with each of these building blocks. We start with ethylene not only because it has the simplest structure but also because it is the most important in terms of tonnage.

In 1977 the United States produced about 25 billion lb of ethylene. This in turn (Sec. 1.5) was converted to 110 billion lb of chemicals and polymers. This figure includes the starting ethylene itself. When we consider that the total US annual production of organic chemicals and polymers was about 250 billion lb we can gauge the importance of ethylene as a raw material. By 1980 it is predicted (see notes) that US capacity for ethylene production will be 40 billion lb.

To be sure, a new process for acetic acid production from methanol and carbon monoxide (Sec. 2.4.3) circumvents the use of ethylene. Similarly, a process is under development that converts CO and H_2 to ethylene glycol (Sec. 2.4.5). These developments however have economic motivation in their own right and are independent of the fortunes of ethylene.

About 60% of ethylene production is destined for polymers. The chemistry of ethylene is shown in Figure 2.5. It falls into three categories. The first [a(1), a(2), and b] involves the polymerization and the oligomerization of ethylene. Reactions a(1) and a(2) indicate that there are two kinds of polyethylene. Low density polyethylene is made by high pressure polymerization (about 1200 atm and 200°C) with oxygen or a peroxide catalyst and a free radical mechanism. Union Carbide has recently announced a new process for low density polyethylene involving milder conditions and a 75% saving of energy costs.

High density polyethylene is made at low pressures of 1–200 atm. either with a metal alkyl catalyst (Ziegler catalysis, Sec. 4.3.10) or a metal oxide supported on silica or alumina (Phillips, Standard Oil of Indiana processes, Sec. 4.3.11).

Unlike the Ziegler process the metal oxide process cannot be applied to propylene and is of less theoretical interest. On the other hand, more high density polyethylene is made by this route than by any other.

Related to Ziegler polymerization is the oligomerization of ethylene as shown in Figure 2.6. Aluminum triethyl reacts with ethylene to form a trialkyl aluminum. The molecular weight distribution of the chains can to some extent be controlled. An aluminum alkoxide results from the interaction of the aluminum trialkyl with oxygen. It can be hydrolyzed to straight chain fatty alcohols with an even number of carbon atoms. Alternatively the aluminum trialkyl may be heated to 280–300°C in the presence of ethylene to give linear α-olefins. Both groups of compounds are important in the manufacture of biodegradable detergents useful in low phosphate-containing formulations (Part II, Sec. 7.3.1). The α-olefins may in turn be oligomerized to trimers to obtain compositions valuable as synthetic lubricants (Part II, Chap. 10).

a
(1) Oxygen, peroxide catalyst → low density polyethylene
(2) Metal alkyl or metal oxide catalyst → high density polyethylene

b → Oligomerization (see Fig. 2.6)

POLYMERIZATION AND OLIGOMERIZATION

c $\xrightarrow{Cl_2}$ CH_2ClCH_2Cl $\xrightarrow{-HCl}$ $CH_2{=}CHCl$
Ethylene dichloride   Vinyl chloride

d $\xrightarrow[\text{HCl, O}_2,\ 250-315°C]{\text{CuCl}_2,\ \text{KCl, Al}_2\text{O}_3\ \text{or SiO}_2}$ CH_2ClCH_2Cl $\xrightarrow{-HCl}$ $CH_2{=}CHCl$

e $\xrightarrow{O_2,\ CH_3COOH,\ PdCl_2/CuCl_2}$ $CH_3CO_2CH{=}CH_2$
Vinyl acetate

f $\xrightarrow[\text{H}_2\text{O}]{O_2,\ \text{PdCl}_2/\text{CuCl}_2/\text{HCl}}$ CH_3CHO
Acetaldehyde

g $\xrightarrow[\text{CH}_3\text{COOH, O}_2]{\text{TeO}_2,\ \text{RX}}$ $CH_3{-}\overset{O}{\overset{\|}{C}}{-}O{-}CH_2{-}CH_2{-}O{-}\overset{O}{\overset{\|}{C}}{-}CH_3$ $\xrightarrow{2H_2O}$
Ethylene glycol diacetate

$2CH_3COOH + HOCH_2CH_2OH$

REPLACEMENT OF ONE OR MORE ETHYLENIC HYDROGENS

$CH_2{=}CH_2 \rightarrow$

h $\xrightarrow{C_6H_6}$ $C_6H_5CH_2CH_3$ $\xrightarrow{-H_2}$ $C_6H_5CH{=}CH_2$
Styrene

i $\xrightarrow{O_2}$ $CH_2{-}CH_2$ (epoxide, O) $\xrightarrow{H_2O}$ $HOCH_2CH_2OH$
Ethylene glycol

j $\xrightarrow{H_2O}$ C_2H_5OH
Ethanol

k $\xrightarrow[\text{CO}]{H_2}$ $CH_3{-}CH_2{-}CHO$
Propionaldehyde
→ $CH_3{-}CH_2{-}COOH$ Propionic acid
→ $CH_3{-}CH_2{-}CH_2OH$ n-Propyl alcohol

l $\xrightarrow{HCl}$ C_2H_5Cl
Ethyl chloride

"CLASSICAL" REACTIONS

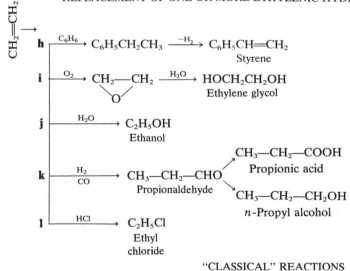

Figure 2.5 Some chemicals from ethylene.

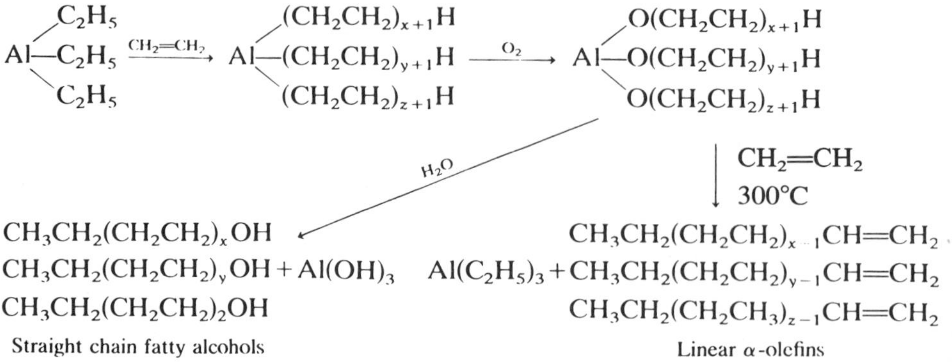

$$Al{\left\langle{{}^{C_2H_5}_{C_2H_5}}_{C_2H_5}\right.} \xrightarrow{CH_2=CH_2} Al{\left\langle{{(CH_2CH_2)_{x+1}H}\atop{(CH_2CH_2)_{y+1}H}}\atop{(CH_2CH_2)_{z+1}H}\right.} \xrightarrow{O_2} Al{\left\langle{{O(CH_2CH_2)_{x+1}H}\atop{O(CH_2CH_2)_{y+1}H}}\atop{O(CH_2CH_2)_{z+1}H}\right.}$$

$$\Big\downarrow{{CH_2=CH_2}\atop{300°C}}$$

H_2O

$CH_3CH_2(CH_2CH_2)_xOH$

$CH_3CH_2(CH_2CH_2)_yOH + Al(OH)_3$ $Al(C_2H_5)_3 + CH_3CH_2(CH_2CH_2)_{x-1}CH=CH_2$

$CH_3CH_2(CH_2CH_2)_2OH$ $CH_3CH_2(CH_2CH_2)_{y-1}CH=CH_2$

Straight chain fatty alcohols $CH_3CH_2(CH_2CH_3)_{z-1}CH=CH_2$

Linear α-olefins

Figure 2.6   Oligomerization of ethylene.

Reactions **c**, **d**, **e**, and **f** of Figure 2.5 involve the substitution of one of ethylene's hydrogens by a functional group—chlorine in **c** and **d**, OH in **f** (in the sense that acetaldehyde is isomeric with unstable vinyl alcohol), and acetoxy in **e**. In **g** two molecules of acetic acid add to yield ethylene glycol diacetate which can be hydrolyzed readily to ethylene glycol and two moles of acetic acid, which are recycled. That so many industrial organic chemicals are prepared by reactions, involving sophisticated catalysts, that "look wrong" to the traditional organic chemist is one reason why industrial organic chemistry can claim to be a separate discipline.

2.4.1   VINYL CHLORIDE

Vinyl chloride is the monomer for poly(vinyl chloride). Originally it was made by addition of hydrogen chloride to acetylene, $CH{\equiv}CH + HCl \rightarrow CH_2 = CHCl$. Ethylene-based vinyl chloride was first made by production of ethylene dichloride, which, when heated in contact with pumice or charcoal, yielded vinyl chloride and a mole of HCl (**c**, Fig. 2.5). The by-product presented a problem, for hydrogen chloride is highly corrosive and difficult to ship. Unless it can be used on site its value is small. Disposal presents economic problems. Furthermore, in this process half the chlorine (produced by electrolysis, which requires expensive electrical energy) was wasted as hydrogen chloride.

One solution was to react the by-product HCl with acetylene. In this way only half as much of the more expensive acetylene was used. Two separate sets of equipment were required, however, one for the ethylene-based and another for the acetylene-based reaction. As the price of ethylene dropped, furthermore, a wholly ethylene-based process became desirable.

The problem was solved by resurrection of a very old process. In 1858 Deacon had shown that HCl can be oxidized to chlorine by air over a cupric chloride catalyst

$$4HCl + O_2 \xrightarrow{CuCl_2} 2H_2O + 2Cl_2.$$

The reaction was reinvestigated and improved catalysts developed. The snag was that it involved an extra stage in vinyl chloride production. The HCl had to be isolated and oxidized and the chlorine recycled. Instead, a related process was developed called oxychlorination. Here the Deacon chemistry has been incorporated into a one-step reaction in which ethylene, hydrogen chloride, and air are passed over a copper chloride/potassium chloride catalyst to give ethylene dichloride. The HCl is oxidized *in situ*, and the chlorine reacts immediately with the ethylene to give ethylene dichloride, which is isolated and cracked to vinyl chloride and by-product HCl, which is recycled (**d**, Fig. 2.5). In the United States about half the ethylene dichloride produced is made by oxychlorination. Chlorination to ethylene dichloride is used when another operation in the same plant, such as oxychlorination, requires the by-product HCl. There is no evidence that anyone uses the Deacon reaction to convert HCl to chlorine. Its value resides in the fact that it led to oxychlorination.

Oxychlorination is typical of the modern petrochemical processes where the aim is to pass a simple feedstock through a hot tube over an appropriate catalyst, the desired chemical emerging at the other end. Of course the secret lies in the "appropriate" catalyst. The next research aim for vinyl chloride production is to develop a process in which the overall reaction can be accomplished without isolation of the ethylene dichloride intermediate. The reaction conditions, particularly temperatures, for the two steps differ so widely, however, that satisfactory yields may not prove possible.

2.4.2   VINYL ACETATE

Poly(vinyl acetate) is an important polymer that forms the basis of many adhesives and water-based emulsion paints. Until 1967 vinyl acetate was made by addition of acetic acid to acetylene, but by 1975 only 10% of US vinyl acetate was made in this way, and the last British plant closed in 1972. The bulk of vinyl acetate (**e**, Fig. 2.5) is now made by an elegant reaction called the Wacker process in which ethylene, oxygen, and acetic acid react in the presence of a palladium-based catalyst.

Two processes were developed, one homogeneous liquid phase with a palladium chloride-cupric chloride catalyst and the other heterogeneous with gas phase reactants and a carbon-$PdCl_2$-$CuCl_2$,$PdCl_2$-Al_2O_3 or palladium-on-carbon catalyst onto which a trickle of potassium acetate solution flows. Fixed and fluidized bed processes are in use. The liquid phase process ran into severe corrosion problems from hydrogen chloride and glacial acetic acid. These might have been overcome, but there were also serious mass transfer problems because foaming in the reactor prevented sufficient ethylene from dissolving quickly enough. The heterogeneous process is now generally used. The Wacker process is also used for the manufacture of acetaldehyde.

2.4.3 ACETALDEHYDE AND ACETIC ACID

In the United States in 1974 1.5 billion lb of acetaldehyde was manufactured. Its production has decreased markedly in recent years (less than 1 billion lb in 1976) because it is less important as a source of *n*-butanol and acetic acid. In 1974 15% was made by oxidation of ethanol, a process now obsolete. The remainder resulted from the Wacker process in which ethylene and oxygen are bubbled through an acidified $PdCl_2$-$CuCl_2$ solution (**f**, Fig. 2.5). The mechanism involves the formation of a pi-bonded complex between ethylene and palladium chloride that decomposes to acetaldehyde and palladium metal. The cupric chloride oxidizes the palladium back to the chloride and is itself reduced to cuprous chloride. All cuprous salts, however, are unstable in the presence of air and are reoxidized to cupric salts.

This is one of the industrial processes where the role of the catalyst is understood.

$$C_2H_4 + PdCl_2 + 2HCl \longrightarrow \left[Cl_3Pd \longleftarrow \begin{array}{c} H \quad\quad H \\ \diagdown \quad \diagup \\ C \\ \| \\ C \\ \diagup \quad \diagdown \\ H \quad\quad H \end{array} \right]^{-} + 2H^+ + Cl^-$$

$$\downarrow H_2O$$

$$CH_3CHO + Pd + 4HCl$$

$$Pd + 2CuCl_2 \longrightarrow PdCl_2 + 2CuCl$$

$$4CuCl + O_2 + 4HCl \longrightarrow 4CuCl_2 + 2H_2O$$

a $2CH_3CHO \longrightarrow CH_3-\underset{\underset{OH}{|}}{CH}-CH_2-CHO \xrightarrow[+H_2]{-H_2O} C_4H_9OH$

n-Butyl alcohol

b $CH_3CHO \xrightarrow[\substack{Mn \text{ or } Co \\ acetate}]{O_2} CH_3COOH \longrightarrow (CH_3CO)_2O$

Acetic acid   Acetic anhydride

c $CH_3CHO + 4HCHO \xrightarrow[H_2O]{alkali} HOH_2C-\underset{\underset{CH_2OH}{|}}{\overset{\overset{CH_2OH}{|}}{C}}-CH_2OH + HCOOH$

Pentaerythritol

d $CH_3CHO \xrightarrow{Cl_2} Cl_3C-CHO$

Trichloroacetaldehyde

Figure 2.7 Reactions of acetaldehyde.

Figure 2.7 shows some of the chemistry depending on acetaldehyde. The aldol condensation in **a** that leads to n-butanol was traditionally important in the chemical industry. n-Butanol is an important solvent and is also used in a range of butyl esters, particularly the acrylate and the methacrylate which are monomers for polymers, and the phthalate which is a plasticizer. The acetate and glycol esters or ethers are used as solvents. The acetaldehyde-based route is being displaced however by the more economical oxo reaction (Sec. 2.12) between propylene, carbon monoxide, and hydrogen.

Acetic acid and acetic anhydride (**b**, Fig. 2.7) are two other important chemicals that can be derived from acetaldehyde. They are used to make a range of acetate esters including vinyl acetate and cellulose acetate. The anhydride is also used in aspirin, though the tonnage is small.

Acetic anhydride is made by two processes. In one, acetic acid (or acetone, but this is uneconomical) is pyrolyzed to ketene which in turn reacts with acetic acid.

$$CH_3COOH \xrightarrow{heat} CH_2{=}C{=}O + H_2O$$

Acetic acid   Ketene

$\xrightarrow{CH_3COOH} (CH_3CO)_2O$

Acetic anhydride

The second procedure involves the *in situ* production from acetaldehyde of peracetic acid, which in turn reacts with more acetaldehyde to yield the anhydride.

$$CH_3CHO + O_2 \longrightarrow CH_3COOOH$$

Acetaldehyde   Peracetic acid

$$\xrightarrow{\quad CH_3CHO \quad} (CH_3CO)_2O + H_2O$$

Acetic acid can also be made by a number of routes. All the acetic acid made in the United Kingdom and an increasing proportion of that made in Japan and Western Europe is based on liquid phase, air oxidation of a paraffinic light naphtha in the C_4–C_5 range. Acetic acid is the main product together with small amounts of formic, propionic, and succinic acids. In the United States where light naphtha is expensive this route is unattractive, but a similar process involving the catalytic liquid phase oxidation of butane from refineries (Sec. 2.6) has until now accounted for about 45% of total production, most of the remainder coming from the oxidation of acetaldehyde. Both of these processes, however, have been challenged by a simple process involving the reaction of carbon monoxide with methanol.

$$CH_3OH + CO \xrightarrow[\text{33–65 atm } I_2\text{-Rh}]{\text{150–200°C}} CH_3COOH$$

The commercialization of this process in the United States has contributed also to the decline of acetaldehyde production. The reaction takes place in the liquid phase and the catalyst is based on iodine-promoted rhodium. A variation has been developed in Germany with the same reactants, a higher temperature (250°C) and pressure (650 atm), and a cobalt iodide catalyst, but it gives lower yields. The advantage of the US process is that yields with respect to methanol are above 99%. Purification is simple, and there is substantial saving on energy intensive separation processes. The first UK plant is due in 1979.

The various methods underline an important characteristic of the chemical industry—there are frequently numerous ways to produce a given chemical, many of which may well be competitive. Furthermore, as more and more is learned about heterogeneous catalysis, the methods involve fewer and fewer stages and bear less and less resemblance to classical organic chemistry.

Pentaerythritol (c, Fig. 2.7) results from the condensation of acetaldehyde with formaldehyde and is a polyhydric alcohol useful in alkyd resins and rigid polyurethane foams. US production of pentaerythritol in 1976 was about 100 million lb.

Chlorination of acetaldehyde (d, Fig. 2.7) leads to trichloroacetaldehyde (chloral) which is condensed with chlorobenzene to give the insecticide DDT now banned in the United States but exported.

2.4.4 STYRENE

After polyethylene and polyvinyl chloride the third large tonnage polymer to be made from ethylene is polystyrene. Styrene provides an example of a "classical" reaction for ethylene. It was used initially not in polystyrene but in a copolymer with butadiene that served as a substitute for natural rubber. Germany had started commercial production of Buna S rubber in 1938 and during World War II was totally dependent on it. With the capture by Japan of the rubber producing areas of Southeast Asia the United States too was forced into a crash program for synthetic rubber. The Buna S-type (GR-S) rubber was the result of an admirable coordinated effort between many academic, industrial, and government laboratories. If either side in the war had lacked synthetic rubber its war effort might have collapsed. Synthetic rubbers now account for about two-thirds of world consumption. Styrene-butadiene (SBR) rubber, the descendant of Buna S and GR-S, is the preferred material for automobile tires (Part II, Sec. 4.3) and accounts for some 60% of world synthetic rubber production, although natural rubber and "synthetic" natural rubber (Part II, Sec. 4.1) are assuming increased importance. Styrene is still widely used in copolymers, but homopolymers or only slightly modified styrene polymers now make up the greater part of the market.

Styrene's precursor is ethylbenzene produced by alkylation of benzene with ethylene over a Friedel-Crafts catalyst such as aluminum chloride. The dehydrogenation stage is carried out over a chromic oxide-supported iron oxide catalyst (**h**, Fig. 2.5).

In a new process styrene results as a coproduct with propylene oxide.

Ethylbenzene is oxidized in the liquid phase to ethylbenzene hydroperoxide. Reaction of propylene with the hydroperoxide provides the desired propylene oxide and phenylmethylcarbinol, which on dehydration yields styrene. This dehydration is a "cleaner," less energy dependent reaction than the dehydrogenation of ethylbenzene. The stoichiometry of the process is such that about 2.5 lb of styrene are produced for every pound of propylene oxide. A new US plant has a capacity of 400 million lb per year of propylene oxide and correspondingly of 1 billion lb of coproduct styrene. Since total styrene production in the United States was 6.3 billion lb per year in 1976, the new plant will provide a sizeable part of what is needed. If its economics meet expectations the profitability of a number of plants required to produce styrene by the conventional route will decrease.

This reaction is interesting for it illustrates a trend towards technology in which two desired products result from one series of reactions. We encounter this again in the cumene hydroperoxide process for phenol and acetone production (Sec. 2.5.3). A possible future route to styrene is described in Section 2.8.

Polyethylene, poly(vinyl chloride), polystyrene, and poly(ethylene terephthalate) (Part II, Sec. 3.9) are the main polymers based on ethylene. Production figures are given in Table 2.5. An indication of the chemical industry's growth problem of the mid-70s is that only with poly(ethylene terephthalate) and high density polyethylene was there growth between 1974 and 1976. Resumption of growth started in 1977.

Table 2.5   Production of Polymers from Ethylene, 1976

	Million lb		
Polymer	United States	Western Europe[a]	Japan
Polyethylene, low density	6470[b]	6600	1900
Poly(vinyl chloride) and copolymers	6250[b]	7200	3200
Polystyrene and copolymers	4630[b]	4300	1820
Poly(ethylene terephthalate) fiber	3640[b]	1800	1100
Polyethylene, high density	3650[b]	2500	900
Styrene-butadiene elastomers	3130[b]	—	—
Poly(vinyl acetate)	526[c]	—	—

[a] Includes the United Kingdom.
[b] 1977 figure.
[c] 1975 figure.

2.4.5 ETHYLENE OXIDE AND GLYCOL

The most important ethylene-based chemical that is not primarily a polymer precursor is ethylene oxide. It is made (**i**, Fig. 2.5) by direct reaction of ethylene and oxygen over a silver catalyst. The reaction is exothermic, and a simultaneous even more exothermic reaction leads to carbon dioxide and water. There is always danger that the reaction will get out of hand, and efficient temperature control is essential.

Ethylene oxide is used in the production of nonionic detergents, ethanolamines for gas scrubbing in refineries, and polyethylene glycols. At least 60% of all ethylene oxide, however, is hydrolyzed to ethylene glycol which has two main uses. One is as an antifreeze in car radiators. The other is for reaction with terephthalic acid to give poly(ethylene terephthalate), which can be spun into polyester fibers.

$$\underset{\text{Ethylene oxide}}{\overset{O}{CH_2-CH_2}} \xrightarrow{H_2O} \underset{\text{Ethylene glycol}}{HOCH_2CH_2OH} \quad \xrightarrow{\underset{\text{Terephthalic acid}}{HOOC-\text{(ring)}-COOH}}$$

$$HOOC-\text{(ring)}-\overset{O}{\underset{\|}{C}}\left[-OCH_2CH_2O-\overset{O}{\underset{\|}{C}}-\text{(ring)}-\overset{O}{\underset{\|}{C}}\right]_n OH$$

Poly(ethylene terephthalate)

The snag in the manufacture of ethylene glycol by way of the direct oxidation of ethylene to ethylene oxide is that overall yields are low—70% is excellent—whereas in many other industrial chemical processes the norm is above 90%. An obsolete route in which ethylene oxide was made by way of ethylene chlorohydrin gave much higher yields but was wasteful of chlorine.

$$CH_2{=}CH_2 + HOCl \text{ (from } Cl_2 + H_2O) \longrightarrow \underset{\underset{OH\quad Cl}{|\quad\ |}}{CH_2-CH_2} \xrightarrow{Ca(OH)_2}$$

$$\overset{O}{CH_2-CH_2} + CaCl_2 + H_2O$$

In a new reaction (**g**, Fig. 2.5) ethylene, acetic acid, and oxygen combine to give ethylene glycol diacetate. These are the reactants that give vinyl acetate in the Wacker Process, but the proportion of acetic acid is greater, and a different catalyst, tellurium oxide plus an alkyl halide, is used. Two ethylenic hydrogen atoms instead of one are substituted. The diacetate can be hydrolyzed to ethylene glycol and the acetic acid recycled so that the overall reaction involves only ethylene, air, and water. An 800 million lb per year plant was scheduled to go on stream in the United States in 1977 but has been plagued by corrosion problems. The market for ethylene oxide could easily be halved.

The acetate process is in turn challenged by an even newer process, which has not yet been commercialized, based on combining CO and H_2 in the presence of a complex rhodium carbonyl catalyst. Ethylene glycol is the predominant product, but there are by-products including propylene glycol, glycerol, and methanol.

2.4.6   ETHANOL, PROPIONALDEHYDE, AND ETHYL CHLORIDE

Industrial ethanol (**j**, Fig. 2.5), once made by fermentation and as such the cornerstone of the aliphatic organic chemicals industry, is now made by hydration of ethylene over a phosphoric-acid-on-celite catalyst. It has humble uses as a basis for ethyl esters, as a solvent, in toiletries and cosmetics, and in a variety of minor applications.

The oxo reaction (Sec. 2.12) with ethylene gives propionaldehyde (**k**, Fig. 2.5), and the addition of HCl across ethylene's double bond gives ethyl chloride (**l**, Fig. 2.5). Propionaldehyde may be reduced to the solvent, *n*-propyl alcohol, or oxidized to propionic acid whose sodium and calcium salts are "rope" inhibitors in bread and in general are used as mold inhibitors and fungicides in foods. An alternative route to *n*-propanol involves the direct oxidation of propane or butane.

Ethyl chloride was once made by reacting ethanol with hydrogen chloride. Today practically all of it is made by the liquid phase addition of HCl to ethylene with an $AlCl_3$ catalyst. Ethyl chloride has specialized uses as a refrigerant, an anesthetic, and in the production of ethyl cellulose, but at least 80% of it has been used in the United States for the production of tetraethyl lead (from ethyl chloride plus a lead-sodium alloy). This usage and correspondingly the production of ethyl chloride are decreasing markedly because tetraethyl lead is being eliminated from gasoline.

Volumes of chemicals produced in the United States from ethylene are shown in Table 2.6.

Table 2.6 Chemicals from Ethylene—Summary

Chemical	Thousand lb 1977
1. Ethylene dichloride	10,480
2. Ethylbenzene	7,300
3. Styrene	6,820
4. Vinyl chloride	5,810
5. Ethylene oxide	4,420
6. Ethylene glycol	3,470
7. Acetic acid[a]	2,580
8. Propionaldehyde	1,950[b]
9. Vinyl acetate	1,600
10. Acetic anhydride[a]	1,500
11. Ethanol	1,300
12. Acetaldehyde	970[b]

[a] Acetic acid and acetic anhydride are also made by direct hydrocarbon oxidation and reaction of CH_3OH and CO.
[b] 1976 figure.

There is more industrial chemistry involving ethylene, but the preceding pages cover the large and medium tonnage uses. Figure 2.8 shows some of the other compounds for which ethylene is the feedstock.

2.5 CHEMICALS FROM PROPYLENE

After ethylene the most important aliphatic raw material is propylene. In 1977 the United States produced about 25 billion lb of propylene, about one-fourth as a coproduct of ethylene production in the thermal cracking of ethane and propane and three-fourths as a by-product of catalytic cracking and of other refinery processes. Half of this amount, 12.6 billion lb, was used for chemicals. Most of the remainder was reacted with isobutane to give alkylates for gasoline. In this respect propylene differs from ethylene, which has few nonchemical uses. Figure 2.9 compares production of various olefins for chemical applications.

Natural gas typically contains only 40% as much propane as ethane. On the other hand, catalytic cracking of higher petroleum fractions which gives negligible amounts of ethylene is a rich source of propylene. Furthermore, steam cracking of ethane, propane, butane, naphtha, or gas

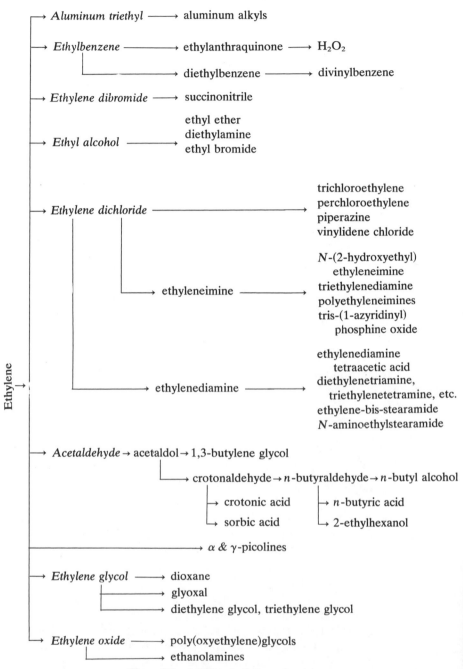

Figure 2.8 Additional ethylene chemistry.

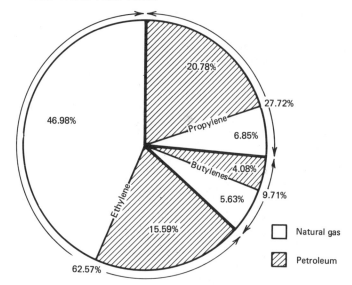

Figure 2.9 Consumption of various olefins.

oil inevitably gives propylene as one of the coproducts. The amount depends on the severity of cracking and for naphtha and gas oil varies from 0.9–0.4 moles propylene mole^{-1} of ethylene as the cracking becomes more severe. A typical figure is about 0.5. In Western Europe and Japan light naphtha cracking is the major source of propylene.

For many years propylene was ethylene's ugly sister, a by-product that was sold at a fraction above its fuel value or else was burned. This availability stimulated the chemist's ingenuity, and many chemical uses have been developed. Propylene's price is still lower than ethylene's although now appreciably above its fuel value.

The main polymers and chemicals made from propylene are shown in Figure 2.10. The propylene equivalent of polyethylene is polypropylene, but it cannot be made by the high pressure free radical process used for low density polyethylene because of the reactivity of propylene's allylic hydrogen atoms. Instead a process (Sec. 4.3.10) analogous to that for high density polyethylene is used (**a**, Fig. 2.10).

The oligomerization of propylene to olefinic dimers, trimers, and tetramers (**b**, Fig. 2.10) takes place in the presence of Friedel-Crafts catalysts such as sulfuric acid. A gasoline fraction results (Sec. 2.2.3) from which the dimers, trimers, and tetramers can be isolated by distillation. The highest volume product is the tetramer known as dodecene. It consists of a mixture of the many possible isomers, and its largest use was

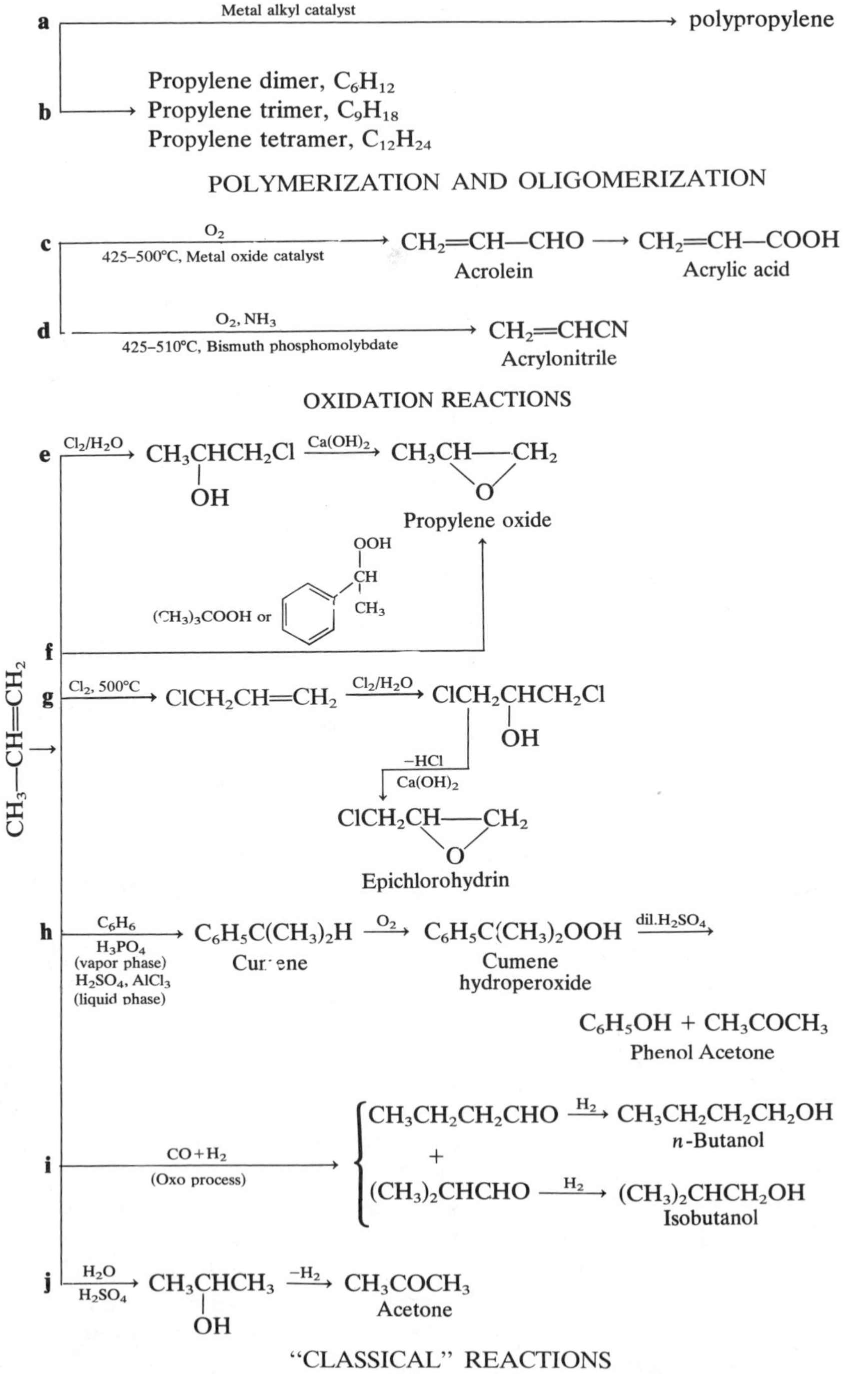

$$CH_3-CH=CH_2 \rightarrow$$

a — $\xrightarrow{\text{Metal alkyl catalyst}}$ polypropylene

b $\longrightarrow$ Propylene dimer, C_6H_{12}
Propylene trimer, C_9H_{18}
Propylene tetramer, $C_{12}H_{24}$

POLYMERIZATION AND OLIGOMERIZATION

c $\xrightarrow[\text{425–500°C, Metal oxide catalyst}]{O_2}$ $CH_2=CH-CHO \rightarrow CH_2=CH-COOH$
Acrolein   Acrylic acid

d $\xrightarrow[\text{425–510°C, Bismuth phosphomolybdate}]{O_2,\ NH_3}$ $CH_2=CHCN$
Acrylonitrile

OXIDATION REACTIONS

e $\xrightarrow{Cl_2/H_2O}$ CH_3CHCH_2Cl $\xrightarrow{Ca(OH)_2}$ $CH_3CH-\!-\!CH_2$
                                $\underset{OH}{|}$ $\underset{O}{\diagdown\!\diagup}$
Propylene oxide

OOH
|
CH
($CH_3)_3COOH$ or [phenyl] CH$_3$

f

g $\xrightarrow{Cl_2,\ 500°C}$ $ClCH_2CH=CH_2$ $\xrightarrow{Cl_2/H_2O}$ $ClCH_2CHCH_2Cl$
                                                        $\underset{OH}{|}$

$\downarrow$ $\overset{-HCl}{\underset{Ca(OH)_2}{}}$

$ClCH_2CH-\!-\!CH_2$
$\underset{O}{\diagdown\!\diagup}$
Epichlorohydrin

h $\xrightarrow[\substack{H_3PO_4 \\ \text{(vapor phase)} \\ H_2SO_4,\ AlCl_3 \\ \text{(liquid phase)}}]{C_6H_6}$ $C_6H_5C(CH_3)_2H$ $\xrightarrow{O_2}$ $C_6H_5C(CH_3)_2OOH$ $\xrightarrow{dil.H_2SO_4}$
Cumene   Cumene
hydroperoxide

$C_6H_5OH + CH_3COCH_3$
Phenol   Acetone

i $\xrightarrow[\text{(Oxo process)}]{CO+H_2}$ $\begin{cases} CH_3CH_2CH_2CHO \xrightarrow{H_2} CH_3CH_2CH_2CH_2OH \\ \qquad\qquad\qquad\qquad\qquad n\text{-Butanol} \\ + \\ (CH_3)_2CHCHO \xrightarrow{H_2} (CH_3)_2CHCH_2OH \\ \qquad\qquad\qquad\qquad\quad \text{Isobutanol} \end{cases}$

j $\xrightarrow[H_2SO_4]{H_2O}$ CH_3CHCH_3 $\xrightarrow{-H_2}$ CH_3COCH_3
                $\underset{OH}{|}$ Acetone

"CLASSICAL" REACTIONS

Figure 2.10 Chemicals and polymers from propylene.

to make dodecylbenzene which was subsequently sulfonated to provide a nonbiodegradable detergent (Part II, Sec. 7.6). This application is now insignificant as more biodegradable detergents have taken over. A smaller use involves alkylation of phenol to provide dodecylphenol, which may then be reacted with ethylene oxide to provide a nonionic detergent. The barium salt of dodecylphenol is used in lubricating oil additives (Part II, Chap. 10). A small amount of the tetramer is subjected to the oxo reaction followed by hydrogenation to provide the tridecyl alcohol useful in plasticizers (Part II, Sec. 11.2). The trimer, by the same series of reactions, yields isodecyl alcohol, which is used more widely in plasticizers than tridecyl alcohol. Nonylphenol is produced in relatively small quantities, and reaction with ethylene oxide converts it to a nonionic detergent. It is also used in lube oil additives (Part II, Chap. 10). Propylene dimer finds little commercial use. It was intended for a new polymer, poly-4-methyl-1-pentene, discovered by Imperial Chemical Industries (ICI) in Britain. Its prospects seemed less promising than ICI had at one time hoped, and the process was sold to a Japanese company where further development continues.

In discussing ethylene's reactions we stressed the unconventional chemistry in which one hydrogen atom was replaced by the functional groups, chlorine, hydroxyl, and acetoxy. This chemistry does not apply to propylene which instead participates in some unconventional chemistry of its own based on its three allylic hydrogen atoms. The greater reactivity of these hydrogens has long been employed in the chlorination reaction (**g**, Fig. 2.10), which yields allyl chloride. This reaction was known for many years before it was discovered that the methyl group of propylene could be oxidized in very much the same way as the methyl group of toluene.

2.5.1 ACRYLIC ACID

As **c** in Figure 2.10 and **f** in Figure 2.11, indicate, propylene may be oxidized directly, first to acrolein and then to acrylic acid. Acrylic acid is a raw material for an important range of esters, methyl, ethyl, butyl, and 2-ethylhexyl acrylates, which are used in the form of emulsion polymers or copolymers as coatings, for paper treatment (giving wet strength), leather finishing, adhesives in nonwoven fabrics, and processing aids for methacrylates.

Acrylic acid and its esters provide another example of the fact that in the chemical industry there are often many competitive ways of arriving at the same product. Figure 2.11 shows six ways to make acrylic acid. Acrylonitrile hydrolysis (**a**) is an obvious one. Another is reaction of HCN

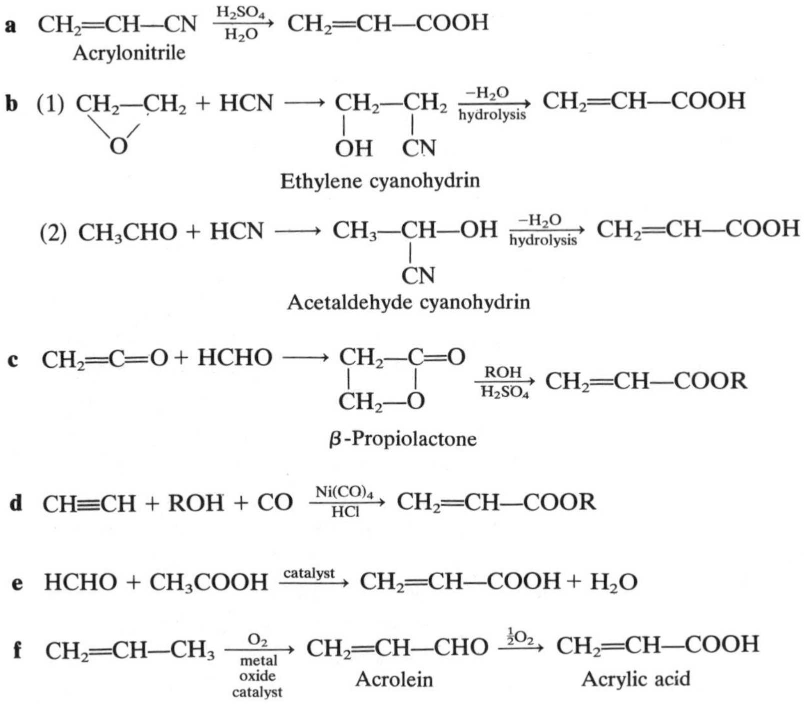

a $CH_2=CH-CN \xrightarrow[H_2O]{H_2SO_4} CH_2=CH-COOH$
  Acrylonitrile

b (1) $CH_2-CH_2 + HCN \longrightarrow CH_2-CH_2 \xrightarrow[hydrolysis]{-H_2O} CH_2=CH-COOH$
  with O bridge, then OH CN
  Ethylene cyanohydrin

  (2) $CH_3CHO + HCN \longrightarrow CH_3-CH-OH \xrightarrow[hydrolysis]{-H_2O} CH_2=CH-COOH$
  with CN
  Acetaldehyde cyanohydrin

c $CH_2=C=O + HCHO \longrightarrow$ $CH_2-C=O$ / CH_2-O $\xrightarrow[H_2SO_4]{ROH} CH_2=CH-COOR$
  β-Propiolactone

d $CH\equiv CH + ROH + CO \xrightarrow[HCl]{Ni(CO)_4} CH_2=CH-COOR$

e $HCHO + CH_3COOH \xrightarrow{catalyst} CH_2=CH-COOH + H_2O$

f $CH_2=CH-CH_3 \xrightarrow[\substack{metal \\ oxide \\ catalyst}]{O_2} CH_2=CH-CHO \xrightarrow{\frac{1}{2}O_2} CH_2=CH-COOH$
  Acrolein   Acrylic acid

Figure 2.11 Acrylic acid—possible production methods.

with ethylene oxide to give ethylene cyanohydrin, which may be simultaneously dehydrated and hydrolyzed with sulfuric acid [**b**(1)]. A third involves the treatment of acetaldehyde with HCN to give acetaldehyde cyanohydrin which may be similarly dehydrated and hydrolyzed [**b**(2)]. The cyanohydrin processes are obsolete.

Reaction **c** in Figure 2.11 starts with ketene from the pyrolysis of acetic acid or acetone. It reacts with formaldehyde to give β-propiolactone, which on reaction with an alcohol gives an acrylate ester directly. The high toxicity and the carcinogenicity of the lactone make this route unattractive.

The next procedure (**d**, Fig. 2.11), in which acetylene is reacted with an alcohol and carbon monoxide in the presence of a nickel carbonyl catalyst to give an alkyl acrylate directly, is a relic of the great days of acetylene chemistry. The process, although obsolescent, still accounted for a little less than half of the total acrylic acid and ester production in the United States in 1976.

In **e** of Figure 2.11 a recently proposed process is shown in which formaldehyde reacts with acetic acid to give acrylic acid, presumably by way of a hydroxymethyl intermediate. This has not yet been commercialized, and there is doubt as to whether it is more economical than propylene oxidation (**f**) which at present is the preferred route. Although the direct oxidation route is of British origin a plant for its use has never been built in the United Kingdom.

2.5.2 ACRYLONITRILE

Closely related to the direct oxidation of propylene to acrolein is oxidation in the presence of ammonia which gives acrylonitrile (**d**, Fig. 2.10). The process, known as ammoxidation, excited much attention, for there was no previous example of the formation of a C-N bond in this way. A bismuth phosphomolybdate catalyst was initially used in the United States. This was replaced by a catalyst based on depleted uranium, which in turn has been superseded by a proprietary catalyst that has increased considerably the yield of acrylonitrile, boosted reactor throughput, and decreased the yield of coproducts. In the early 1960s ammoxidation displaced various processes such as the addition of HCN to acetylene and the interaction of ethylene oxide with HCN followed by dehydration.

Another commercial process, now obsolete, involved the reaction of propylene with nitric oxide. On the horizon is a process based on the ammoxidation of propane rather than propylene. This is formally analogous to the reaction in which butane can be oxidized to maleic anhydride (Sec. 2.6.2) by the simultaneous introduction of a double bond and functional groups into saturated hydrocarbons.

The fact that the methyl group of propylene can be preferentially oxidized to give acrolein indicates that not all of the hydrogens of propylene are equivalent. The methyl hydrogens are allylic and hence reactive; the allyl radical $CH_2 \text{-----} CH \text{-----} CH_2$, on the other hand, is relatively stable. It was the recognition of this difference that led to the ammoxidation process. Acetonitrile and HCN are by-products of the reaction, and acrylonitrile production is now so large that it is the chief source of these materials. HCN's main use is in the production of methyl methacrylate (Sec. 2.5.3). The second largest use of HCN is for production of adiponitrile (Sec. 2.6.1).

Butadiene-acrylonitrile rubbers were in use before World War II (Part II, Sec. 4.2.5). Polyacrylonitrile fibers were investigated but their commercialization had to await the development of a solvent from which they could be spun. Dimethylformamide[$O{=}CH{-}N(CH_3)_2$] proved suitable, and polyacrylonitrile fibers now account for about 55% of the total acrylonitrile market in the United States and 80% in the United Kingdom. Familiar trade names are Orlon, Acrilan, and Courtelle.

A more recent use for acrylonitrile is its hydrodimerization to adiponitrile, which can be reduced to hexamethylenediamine for nylon production (Part II, Sec. 2.15). Adiponitrile can also be hydrolyzed to adipic acid, but this is not economically feasible.

2.5.3 ACETONE

Ammoxidation is far removed from classical organic chemistry. The cumene peroxidation route to phenol and acetone (**h**, Fig. 2.10), although one of the first true petrochemical processes, is much closer to it. Alkylation of benzene with propylene in the presence of a Friedel-Crafts catalyst gives cumene. Air is bubbled through the cumene, and the tertiary carbon atom which is made even more reactive by the nearby aromatic ring is oxidized to give cumene hydroperoxide. The hydroperoxide is easily decomposed with a small amount of sulfuric acid at 45–65°C to phenol and acetone. The mechanism involves the phenide shift to a positive oxygen atom shown in Figure 2.12. This reaction is a route to both phenol and acetone. As with the propylene oxide-styrene process (Sec. 2.4.4), coproducts are manufactured in one set of equipment.

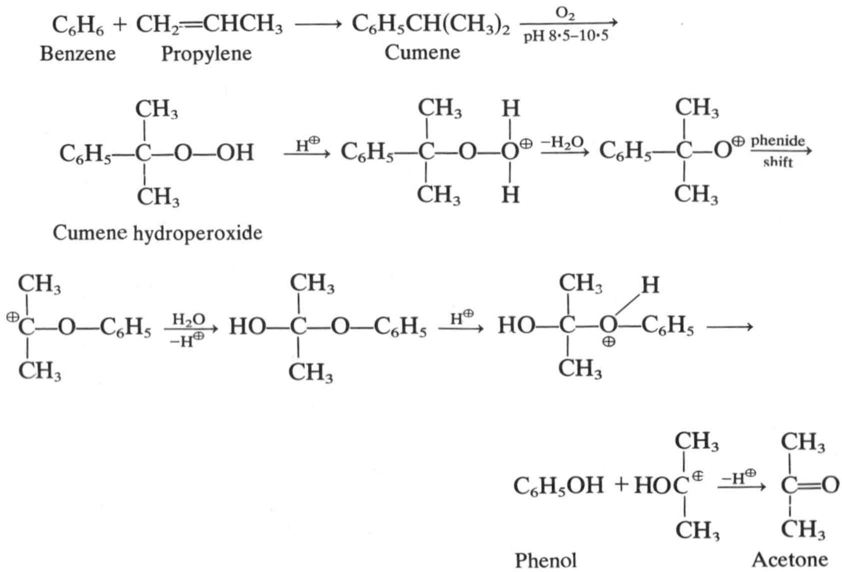

$$C_6H_6 + CH_2{=}CHCH_3 \longrightarrow C_6H_5CH(CH_3)_2 \xrightarrow[pH\ 8\cdot5-10\cdot5]{O_2}$$

Benzene   Propylene   Cumene

Figure 2.12 Mechanism of cumene hydroperoxide decomposition.

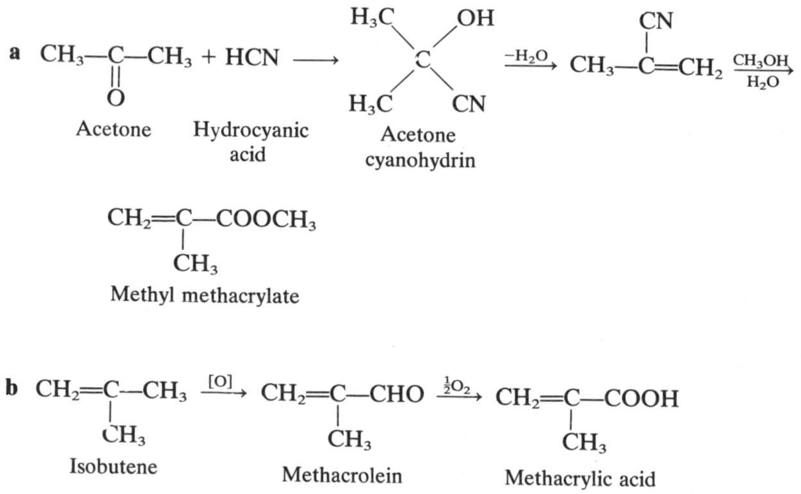

a $CH_3-\underset{\underset{O}{\|}}{C}-CH_3 + HCN \longrightarrow$ $\underset{\underset{H_3C}{}}{\overset{H_3C}{\underset{}{}}}\underset{CN}{\overset{OH}{C}}$ $\xrightarrow{-H_2O}$ $CH_3-\underset{\underset{}{}}{\overset{CN}{C}}=CH_2$ $\xrightarrow[H_2O]{CH_3OH}$

Acetone   Hydrocyanic   Acetone
              acid   cyanohydrin

$CH_2=\underset{\underset{CH_3}{|}}{C}-COOCH_3$

Methyl methacrylate

b $CH_2=\underset{\underset{CH_3}{|}}{C}-CH_3$ $\xrightarrow{[O]}$ $CH_2=\underset{\underset{CH_3}{|}}{C}-CHO$ $\xrightarrow{\frac{1}{2}O_2}$ $CH_2=\underset{\underset{CH_3}{|}}{C}-COOH$

Isobutene   Methacrolein   Methacrylic acid

Figure 2.13   Methyl methacrylate syntheses.

The competitive route to acetone is shown in **j** of Figure 2.10. Absorption of propylene in concentrated sulfuric acid gives isopropyl sulfate, which is hydrolyzed to isopropanol. This has a number of minor uses primarily as a solvent, but most of it is air-oxidized or dehydrogenated to acetone. Given a market for phenol, the cumene peroxidation route is the more economical route to acetone. Any shortfall is then made up by the isopropanol route, which on a world basis probably accounts for 40% of acetone production.

An important use for acetone is as a raw material for methyl methacrylate in the multistage process shown in **a** of Figure 2.13. About 500 million lbs were made in the United States in 1976. The apparently attractive oxidation route from isobutene by way of methacrolein (**b**, Fig. 2.13) using the same chemistry that converts propylene to acrylic acid (Sec. 2.5.1) was tried commercially in the 1960s but was not successful, possibly because isobutene contains two methyl groups, both with allylic hydrogens and both liable to oxidation. Recent announcements from Japan, however, describe two processes under development, both involving air oxidation over newly developed catalysts. Oxidation of t-butanol is the basis for a plant now being constructed in the United States for the manufacture of methyl methacrylate. Apparently, t-butanol oxidizes more smoothly to methacrylic acid than does isobutene. Another route to methyl methacrylate is in a reaction that produces it as a coproduct with propylene oxide as described in the next section. Another use for acetone

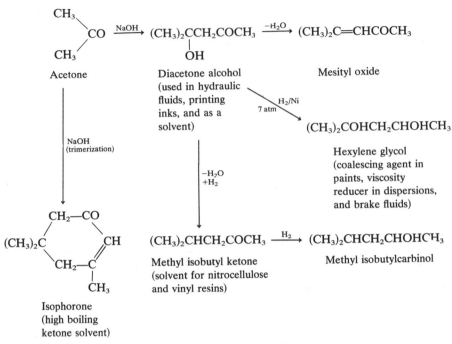

Figure 2.14   Acetone reactions.

is reaction with phenol to give bisphenol A (Sec. 2.7.1). Other uses are shown in Figure 2.14.

2.5.4   PROPYLENE OXIDE

Propylene oxide (**e, f,** Fig. 2.10) was made traditionally by reaction of propylene with hypochlorous acid (generated *in situ* from chlorine and water) followed by dehydrochlorination with calcium hydroxide. The process was at one time used for ethylene oxide, and many of the plants were converted to propylene oxide when the direct oxidation route to ethylene oxide appeared. The process is wasteful of chlorine, and much research was devoted to discovery of a direct oxidation route by way of various hydroperoxides. *t*-Butyl hydroperoxide, from the oxidation of isobutane, was the first to be used commercially. *t*-Butanol is a by-product in both its formation and in the oxidation step.

Approximately 2.2 moles of *t*-butanol mole^{-1} of propylene oxide are produced, and the economics are said to hinge on the market for the by-product. Since *t*-butanol is a white solid (m.p. 25.6°C) and has no bulk

$$2CH_3\overset{\underset{\displaystyle CH_3}{|}}{C}HCH_3 + \tfrac{3}{2}O_2 \longrightarrow CH_3\overset{\underset{\displaystyle OOH}{|}}{\underset{|}{C}}\overset{CH_3}{}CH_3 + H_3C\overset{\underset{\displaystyle CH_3}{|}}{\underset{|}{C}}\overset{CH_3}{}OH$$

t-Butanol
hydroperoxide   t-Butanol

$$\downarrow CH_3CH{=}CH_2$$

$$CH_3CH{-}CH_2 + H_3C\overset{\underset{\displaystyle CH_3}{|}}{\underset{|}{C}}\overset{CH_3}{}OH$$
$$\underset{O}{\diagdown\diagup}$$

Propylene   t-Butanol
oxide

uses, it must be converted to isobutene which is useful for alkylation for gasoline production. If the t-butanol route to methyl methacrylate is a success (Sec. 2.5.3), this situation will change.* Ethylbenzene gives phenyl-methylcarbinol which can be dehydrated to styrene as already described (Sec. 2.4.4). Isopentane gives coproduct isopentanol, which can be converted to isoprene (used in synthetic rubbers), although this is probably not being done commerically.

It has been suggested that a process is being commercialized for the production of propylene oxide and methyl methacrylate using perisobutyric acid, which in turn is obtained from isobutyraldehyde as the following equations indicate:

$$CH_3\overset{\underset{\displaystyle CH_3}{|}}{\underset{|}{C}}\overset{H}{}CHO \xrightarrow{O_2} CH_3\overset{\underset{\displaystyle CH_3}{|}}{\underset{|}{C}}\overset{H}{}COOOH$$

Isobutyraldehyde   Perisobutyric acid

$$CH_3\overset{\underset{\displaystyle CH_3}{|}}{\underset{|}{C}}\overset{H}{}COOOH + CH_2{=}CH{-}CH_3 \longrightarrow CH_2{-}CH{-}CH_3 + $$
$$\underset{O}{\diagdown\diagup}$$

$$CH_3CHCOOH$$
$$\underset{CH_3}{|}$$

Propylene   Propylene
oxide

$$CH_3\overset{\underset{\displaystyle CH_3}{|}}{\underset{|}{C}}\overset{H}{}COOH \xrightarrow[CH_3OH]{-H_2} CH_2{=}\overset{\underset{\displaystyle CH_3}{|}}{C}COOCH_3 + H_2O$$

Isobutyric acid   Methyl methacrylate

* t-Butanol has recently been proposed as an octane improver in gasoline. Methyl t-Butyl ether has now been approved in the U.S.A. as an octane improver.

Isobutyraldehyde has always been an unwanted by-product from the reaction of CO and H_2 with propylene (Sec. 2.5.6). Lack of a market for it motivated the development of an elegant catalyst system for oxo reactions that provides high yields of n-butyraldehyde. The above reactions, if commercial, will make useful the old oxo reaction catalysts.

Propylene oxide usage has increased markedly since 1968 because it is the precursor for the polypropylene glycols useful in urethane formation (Part II, Sec. 2.7). In the United States in 1977 approximately 1.8 billion lb were consumed. The peroxidation route accounted for an estimated 50% of US production.

Propylene oxide is used mainly for the production of polyols and polyethers with hydroxyl end groups.

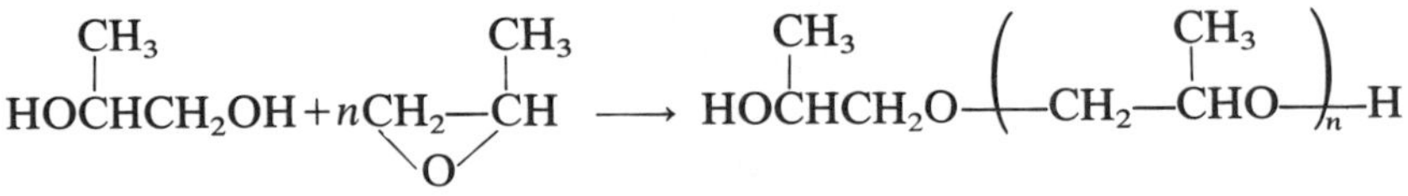

$$\underset{\text{Propylene glycol}}{\text{HOCHCH}_2\text{OH}} + n\underset{}{\text{CH}_2-\text{CH}} \longrightarrow \underset{\text{Poly (propylene glycol)}}{\text{HOCHCH}_2\text{O}-\left(-\text{CH}_2-\text{CHO}-\right)_n-\text{H}}$$

The polymerization proceeds head-to-tail to give secondary hydroxyl end groups. When the polyols and polyethers react with diisocyanates, especially diisocyanotoluene (tolylene diisocyanate, toluene diisocyanate, TDI) to give polyurethane resins, the less reactive secondary hydroxyl groups react more controllably with the isocyanate groups. For some reactions, however, higher reactivity is required, and the secondary hydroxyls are reacted with ethylene oxide to provide primary hydroxyl end groups. If a more highly cross-linked polymer is desired, as is frequently the case, a triol can be used prepared from the interaction of glycerol with three moles of propylene oxide (Sec. 4.4.2).

2.5.5 EPICHLOROHYDRIN

In **g** of Figure 2.10 the production is shown of epichlorohydrin by a series of reactions analogous to the traditional routes to ethylene and propylene oxides. The process is wasteful of chlorine but has not yet been replaced. This is probably due not to the inherent difficulty of finding a new route but to the smaller incentive offered by the lower consumption of epichlorohydrin. The first step of the selective chlorination takes place because the hydrogens on the methyl group of propylene are allylic and hence reactive. Epichlorohydrin reacts with bisphenol A to give epoxy resins (Part II, Sec. 2.3.5) and may also be hydrolyzed by alkali to glycerol. "Natural" glycerol from hydrolysis of fats and oils is still able to compete and held about 43% of the US market in 1977 although prices were depressed by overcapacity.

There is yet a third route to glycerol. Acrolein produced by direct oxidation of propylene (Sec. 2.5.1) may be reduced with isopropanol to allyl alcohol. Acetone is the by-product. Treatment of the allyl alcohol with hydrogen peroxide gives glycerol. The route saves on chlorine but loses on expensive hydrogen peroxide.

2.5.6 BUTYRALDEHYDE AND ISOBUTYRALDEHYDE

Propylene (**i**, Fig. 2.10) reacts with carbon monoxide and hydrogen (Secs. 2.12; 5.5) at 130–175°C and 250 atm over a cobalt carbonyl catalyst to give a mixture of between 3 and 4 moles of *n*-butyraldehyde to 1 mole of isobutyraldehyde. This reaction is called the oxo process or hydro-formylation. Reduction gives *n*-butanol and isobutanol which are used unchanged or as their acetates for lacquer solvents. The straight chain compound has the premium properties, and catalysts are now available (primarily organophosphine ligand-modified cobalt carbonyls, cobalt carbonyl hydride, or rhodium chloride with ligands such as triphenylphosphine) that give high yields of primary aldehydes in contrast to most olefin reactions, which give secondary compounds. For example the oxo reaction with propylene with one of the new catalyst systems may give the alcohol precursors, butyraldehyde and isobutyraldehyde in the ratio of 10:1. As a result the oxo process is now preferred to the aldol synthesis (Secs. 2.4.3; 2.5.6) for preparation of primary alcohols such as *n*-butanol. The phosphine ligands increase the activity of the cobalt carbonyl catalyst. Because they are bulky the stearic requirements of the ligands facilitate anti-Markownikoff addition to yield a higher amount of the normal rather than the iso structure. Another important advantage is that with the new catalyst the reaction takes place at lower temperatures and pressures.

Lower pressure variants of the oxo process are under development, and a single stage process to alcohols that incorportates the hydrogenation step could be important in the future.

Isobutyraldehyde reacts with two moles of formaldehyde to yield neopentyl glycol used largely in unsaturated polyester synthesis to enhance alkali resistance of the cured product. This preparation is analogous to that of pentaerythritol (Sec. 2.4.3).

$$\underset{\substack{|\\ CH_3}}{\overset{\substack{H\\ |}}{CH_3-C-CHO}} + 2HCHO \longrightarrow \underset{\substack{|\\ CH_3}}{\overset{\substack{CH_2OH\\ |}}{CH_3-C-CH_2OH}} + HCOOH$$

Isobutyraldehyde Formaldehyde   Neopentyl glycol   Formic acid *n*-Butyraldehyde will undergo the aldol condensation to provide an intermediate that loses water to give an unsaturated aldehyde. Reduction gives the important plasticizer alcohol, 2-ethylhexanol (Part II, Sec. 11.2).

$$2CH_3CH_2CH_2CHO \longrightarrow CH_3CH_2CH_2\underset{\underset{OH}{|}}{CH}CH(C_2H_5)CHO \xrightarrow{-H_2O}$$

$$CH_3CH_2CH_2CH{=}C(C_2H_5)CHO \xrightarrow{H_2}$$

$$CH_3CH_2CH_2CH_2CH(C_2H_5)CH_2OH$$
2-Ethylhexanol

2.5.7 METATHESIS

As already mentioned, propylene tends to be available in quantities too great for the market and thus sells at a low price. Relative prices are shown in Table 6.1. A process that has excited organic chemists worldwide has been devised whose objective is to allow greater flexibility in olefin production with a consequent increase in profitability. Two moles of propylene react together in the presence of one of many possible catalysts including Lewis acids and salts of tungsten, molybdenum, or rhenium in a so-called metathesis reaction to give 1 mole of 2-butene and one of ethylene.

$$2CH_2{=}CH{-}CH_3 \xrightarrow{catalyst} \begin{bmatrix} CH_2{-}CH{-}CH_3 \\ \text{Cat.} \\ CH_2{-}CH{-}CH_3 \end{bmatrix} \longrightarrow \begin{matrix} CH_2 \\ || \\ CH_2 \end{matrix} + \begin{matrix} CH{-}CH_3 \\ || \\ CH{-}CH_3 \end{matrix}$$

Propylene   "Quasi-cyclobutane"   Ethylene   Butene-2
intermediate

Both of the products sell in the United States for higher prices than the starting material. The process has been operated in Canada on a pilot basis but has not been used commercially. It has surprisingly broad applicability and can be applied to almost any non-conjugated olefin. With oleic acid an unsaturated dibasic acid and unsaturated hydrocarbon result as follows:

$$2CH_3{-}(CH_2)_7{-}CH{=}CH{-}(CH_2)_7{-}COOH \rightarrow$$
$$HOOC{-}(CH_2)_7{-}CH{=}CH{-}(CH_2)_7{-}COOH$$
$$+ CH_3{-}(CH_2)_7{-}CH{=}CH{-}(CH_2)_7{-}CH_3$$

Cyclopentene disproportionates to cyclodecadiene which by continued disproportionation can form oligomers and polymers.

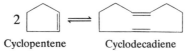

Cyclopentene   Cyclodecadiene

Acetylenes and pseudoolefins such as cyclopropane also undergo metathesis. Of potential commercial interest is the metathesis reaction that takes place between stilbene and ethylene (Sec. 2.8) to yield styrene. If it were ever necessary to use carbohydrates as a source of chemicals, ethylene from dehydration of fermentation ethanol could undergo a metathesis reaction with 2-butene from ethylene dimerization (Sec. 3.4.2) to give propylene. Thus a reaction devised to use excess propylene might eventually be used to prepare it. Metathesis is currently being used in a new process to prepare α-olefins and alcohols as detergent intermediates (see notes, Part II, Sec. 7.3.1).

2.6 CHEMICALS FROM THE C$_4$ STREAM

We have described the huge volume of chemicals based on ethylene and the somewhat smaller volume based on propylene. Quantitatively the C$_4$ stream provides far less chemicals than ethylene or propylene. Only one C$_4$-based chemical, butadiene, is on the list of the 50 chemicals produced in highest volume in the United States. Obviously butylenes are not in the same league as ethylene's 25 billion lb and propylene's 12.6 billion lb. But this is not the whole story. It was the C$_4$ stream that provided the synthetic rubber vital to both sides in World War II. Also the chemistry associated with the C$_4$ stream is interesting and different from what we have discussed so far, and thus we shall deal with it in comparable detail.

Only about 5% of the aliphatics in the natural gas fraction that is cracked comprises n-butane so only a small volume of C$_4$ olefins comes from natural gas. The remainder comes from catalytic cracking of petroleum, steam cracking of naphtha, and from refinery off-gases—the gases formed in refineries as by-products of processes such as catalytic reforming and catalytic cracking. The main components of the C$_4$ stream are n-butane, 1-butene, 2-butene, isobutene, and 1,3-butadiene (Fig. 2.15). These can to some extent be interconverted. n-Butane readily isomerizes to isobutane, and both n-butane and the butenes can be dehydrogenated to 1,3-butadiene.

The usage of the C$_4$ stream illustrates some differences between the United States and Western Europe. In Europe there is virtually no market for alkylate gasoline so there is little difference in value between

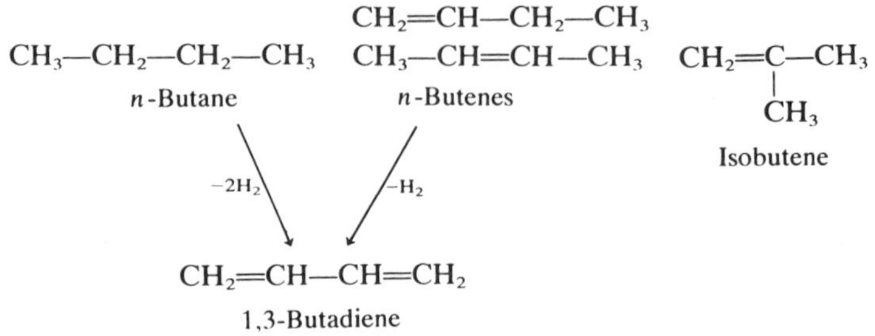

$$CH_2{=}CH{-}CH_2{-}CH_3$$

$$CH_3{-}CH_2{-}CH_2{-}CH_3 \qquad CH_3{-}CH{=}CH{-}CH_3 \qquad CH_2{=}\overset{\displaystyle |}{\underset{\displaystyle CH_3}{C}}{-}CH_3$$

n-Butane   *n*-Butenes   Isobutene

$$CH_2{=}CH{-}CH{=}CH_2$$

1,3-Butadiene

Figure 2.15 Major components of C_4 stream.

butanes and butenes. In the United States, on the other hand, the butenes are expensive because of their value as feedstocks for alkylate gasoline and for chemical synthesis. Isobutane is also used in alkylation but is cheap because it is readily produced from butane. The value of the butenes for chemical synthesis, in turn, hinges on the fact that the refinery mixture of 1- and 2-butenes is in demand for dehydrogenation to 1,3-butadiene for synthetic rubbers. And, to return to our old theme, Western Europe's requirements of 1,3-butadiene can be satisfied by simple extraction from the C_4 stream from the steam cracking of naphtha. As the United States turns more to steam cracking of gas oil we may expect to see her patterns come closer to those of the West Europeans.

2.6.1 CHEMICALS FROM BUTADIENE

Butadiene is the most important C_4 hydrocarbon, and its most important use is in synthetic rubbers. Indeed, half of it goes into styrene-butadiene rubbers (SBR) which are widely used in automobile tires. Also SBR is blended with styrene homopolymers to increase their impact resistance.

Styrene-butadiene copolymers were the original basis for latex paint, but poly(vinyl acetate) and polyacrylate latices have superior properties and have since captured the market. Styrene-butadiene latices, on the other hand, are still important as binders for paper and for carpet backing.

Several other synthetic rubbers are based on butadiene. The homopolymer, polybutadiene rubber, which accounts for about 20% of total usage, is blended with natural and other rubbers such as SBR to aid their processing. It is also used in tire treads to increase wear resistance, low temperature flexibility, and resistance to aging (Part II, Sec. 4.3). Butadiene may also be copolymerized with acrylonitrile to give the specialized nitrile rubbers that have exceptional resistance to organic solvents, and with styrene, and acrylonitrile to give the so-called ABS resins (Part II, Sec. 2.1.3).

The sophistication of modern polymerization techniques is such that it is possible to polymerize a monomer like butadiene to give either elastomers or hard, resinous materials, and the various polybutadienes are discussed in Section 4.3.10.

The chlorination of butadiene in the vapor phase gives a mixture of 3,4-dichloro-1-butene, $CH_2{=}CH{-}CHCl{-}CH_2Cl$, and 1,4-dichloro-2-butene, $ClCH_2{-}CH{=}CH{-}CH_2Cl$. The latter may be isomerized to the former by heat in the presence of cuprous chloride. Dehydrochlorination of 3,4-dichloro-1-butene, leads to "chloroprene," the monomer for neoprene rubbers. This use consumes about 10% of butadiene production.

Neoprene was an early synthetic rubber. Its lack of resilience makes it useless in tires, the major consumer of rubber, but it has good resistance to oil and ozone which suits it for many specialized uses. In the early days it was made by way of acetylene that was dimerized and treated with HCl. Again we see petrochemicals replacing acetylene, for duPont phased out the last acetylene-based polychloroprene plant in the United States in 1974 and in Northern Ireland in 1977.

$$2CH{\equiv}CH \longrightarrow CH_2{=}CH{-}C{\equiv}CH \xrightarrow{\text{HCl}} CH_2{=}CH{-}\underset{\underset{Cl}{|}}{C}{=}CH_2$$

Vinylacetylene   "Chloroprene"

To return to the chemistry of butadiene, addition of sulfur dioxide followed by hydrogenation yields tetramethylene sulfone (Sulfolane) which is used to extract aromatic compounds from hydrocarbon streams in petroleum refineries.

$$CH_2{=}CH{-}CH{=}CH_2 + SO_2 \longrightarrow \underset{\underline{\qquad SO_2 \qquad}}{CH_2{-}CH{=}CH{-}CH_2} \xrightarrow{\text{H}_2}$$

$$\underset{\underline{\qquad SO_2 \qquad}}{CH_2{-}CH_2{-}CH_2{-}CH_2}$$

Cyclization of 2 and 3 moles of butadiene gives 1,5-cyclooctadiene and 1,5,9-cyclododecatriene. The latter is an intermediate for perfumes that can also be hydrogenated to cyclododecane. Treatment with nitrosyl chloride gives laurolactam, the monomer for nylon 12. The same process is used to obtain caprolactam from cyclohexane, the monomer for nylon 6 (Sec. 2.7.2). The triene can also be hydrogenated to cyclododecene whose oxidation with nitric acid gives dodecanedioic acid.

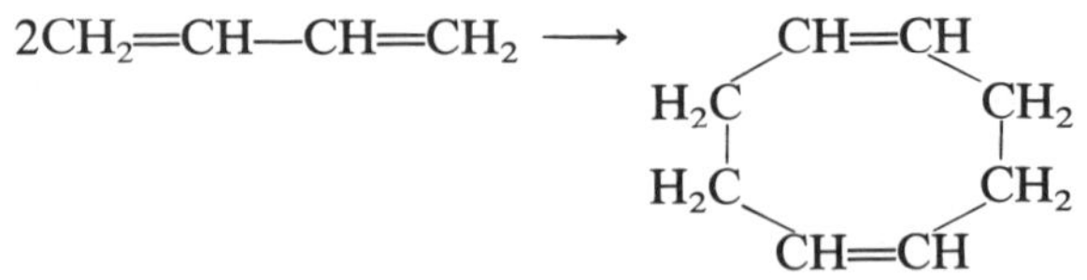

$$2CH_2\!\!=\!\!CH\!\!-\!\!CH\!\!=\!\!CH_2 \longrightarrow$$

1,5-Cyclooctadiene

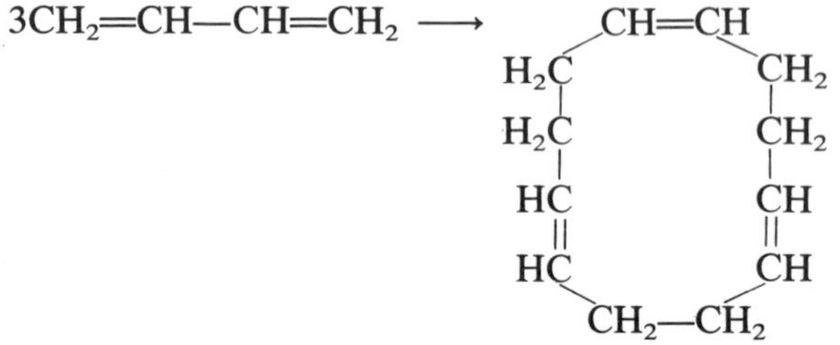

$$3CH_2\!\!=\!\!CH\!\!-\!\!CH\!\!=\!\!CH_2 \longrightarrow$$

1,5,9-Cyclododecatriene

Finally, butadiene is the classic diene for the Diels-Alder reaction. This reaction is used industrially on a relatively small scale. An example is the reaction of butadiene and maleic anhydride to give tetrahydrophthalic anhydride used in polyester and alkyd resins. The Diels-Alder reaction is also used to give ethylidene *nor*-bornene from butadiene and cyclopentadiene.

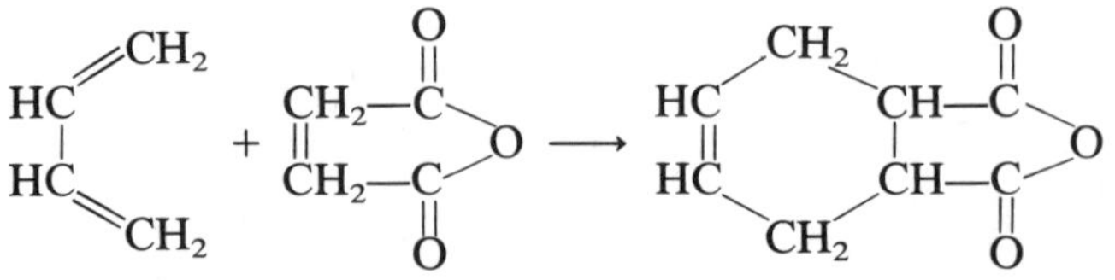

1,3-Butadiene   Maleic anhydride   Tetrahydrophthalic anhydride

Butadiene, although consumed on a scale almost an order of magnitude less than ethylene, has no fewer nor less interesting reactions. They are carried out however on a smaller scale.

2.6.2   HEXAMETHYLENEDIAMINE

Butadiene figures prominently in the most economic synthesis of hexamethylenediamine (HMDA). Indeed the evolution of syntheses for HMDA provides an excellent example of how sophisticated processes develop in the chemical industry. The first route to HMDA involved traditional reactions in which adipic acid (Sec. 2.7.2) and ammonia were reacted to give ammonium adipate which on dehydration first to the amide and then to the nitrile provides adiponitrile, which in turn may be hydrogenated to hexamethylenediamine.

$$HOOC—(CH_2)_4—COOH \xrightarrow{NH_3} [NH_4OOC—(CH_2)_4—COONH_4] \xrightarrow{-H_2O}$$

Adipic acid Ammonium adipate

$$\left[H_2N—\overset{\overset{\displaystyle O}{\|}}{C}—(CH_2)_4—\overset{\overset{\displaystyle O}{\|}}{C}—NH_2 \right] \xrightarrow{-H_2O} N{\equiv}C—(CH_2)_4—C{\equiv}N \xrightarrow{H_2}$$

Adipamide Adipontrile

$$H_2N—(CH_2)_6—NH_2$$

Hexamethylene
diamine

Butadiene entered the picture because of a synthesis in which the 1,4-dichloro-2-butene mentioned earlier was treated with sodium cyanide to give 1,4-dicyano-2-butene. This may be hydrogenated in two steps, first to 1,4-dicyanobutane (adiponitrile) and then to hexamethylenediamine.

$$ClCH_2—CH{=}CH—CH_2Cl \xrightarrow{\substack{NaCN \\ CuCl}} N{\equiv}C—CH_2—CH{=}CH—CH_2—C{\equiv}N$$

1,4-Dicyano-2-butene

$$\xrightarrow{H_2} N{\equiv}C—CH_2—CH_2—CH_2—CH_2—C{\equiv}N \xrightarrow{H_2} H_2N—(CH_2)_6—NH_2$$

1,4-Dicyanobutane Hexamethylenediamine

The electrohydrodimerization of acrylonitrile (Sec. 2.5.2) was the next synthesis. It was an ingenious departure from classical chemistry.

$$N{\equiv}C—CH{=}CH_2 + CH_2{=}CH—C{\equiv}N \xrightarrow[2e]{2H^+} N{\equiv}C—(CH_2)_4—C{\equiv}N$$

Acrylonitrile Adiponitrile

It is rivaled by still another process (Fig. 2.16) which turns out to be the most economic. This process starts with the 1,4 addition of HCN to butadiene. The catalyst (e.g., nickel salt with appropriate ligands) shifts the double bond to a terminal position. This same catalyst makes possible the addition of HCN to the terminal bond in an anti-Markownikoff fashion. If it were not possible to use the same catalyst for both reactions, the double bond shift could not be carried out in high yield because of the achievement of an equilibrium. The resulting adiponitrile may be hydrogenated to hexamethylenediamine. This synthesis, as opposed to the one starting with the chlorination of butadiene, eliminates the need for costly chlorine, uses HCN which is now cheaper than NaCN (but not when the synthesis based on 1,4-dichloro-2-butene was devised), eliminates the handling of

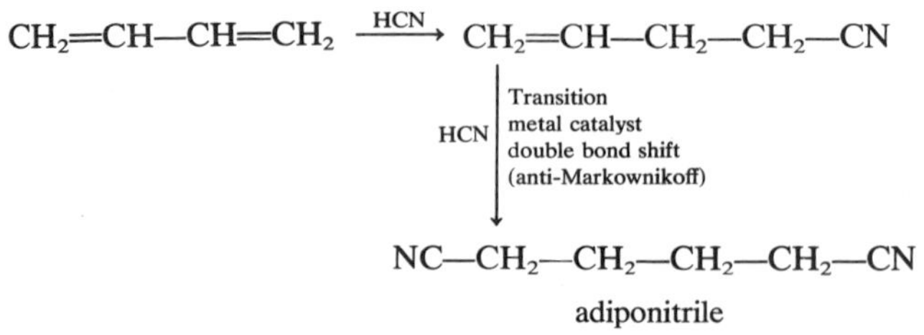

$$CH_2{=}CH{-}CH{=}CH_2 \xrightarrow{\text{HCN}} CH_2{=}CH{-}CH_2{-}CH_2{-}CN$$

HCN | Transition
metal catalyst
double bond shift
(anti-Markownikoff)

$$NC{-}CH_2{-}CH_2{-}CH_2{-}CH_2{-}CN$$
adiponitrile

Figure 2.16 Newest synthesis of adiponitrile.

carcinogenic 1,4-dichloro-2-butene, and solves the problem of disposing of ecologically unacceptable copper-contaminated NaCl.

The mechanism of the HCN addition (Fig. 2.17) involves the formation of an H-M-CN complex in which M is the metallic catalyst. This in turn pi bonds with the terminal olefin after which a sigma-bonded complex forms that subsequently eliminates the metal to provide the adiponitrile. Very important in the synthesis are ligands whose identity has not been

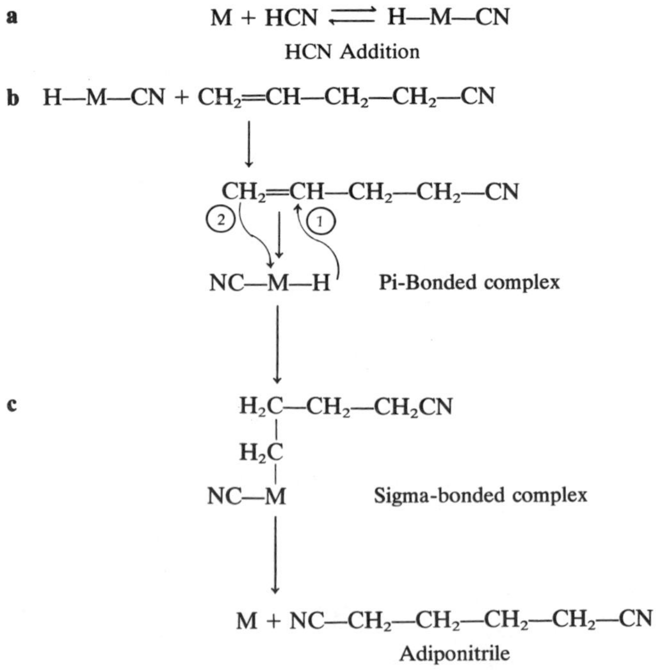

a $M + HCN \rightleftharpoons H{-}M{-}CN$

HCN Addition

b $H{-}M{-}CN + CH_2{=}CH{-}CH_2{-}CH_2{-}CN$

$$CH_2{=}CH{-}CH_2{-}CH_2{-}CN$$
②⟍ ↓ ①↗
$$NC{-}M{-}H$$ Pi-Bonded complex

c
$$H_2C{-}CH_2{-}CH_2CN$$
$$H_2C$$
$$NC{-}M$$ Sigma-bonded complex

$$M + NC{-}CH_2{-}CH_2{-}CH_2{-}CH_2{-}CN$$
Adiponitrile

Ligands omitted.

Figure 2.17 Mechanism of HCN addition in adiponitrile synthesis.

disclosed. That this synthesis is more economical than the acrylonitrile dimerization underlines the role of raw materials and energy costs. Acrylonitrile is considerably more expensive than butadiene, and the electrochemical reactions are energy intensive.

2.6.3 CHEMICALS FROM BUTYLENES

Figure 2.15 shows the C$_4$ stream. Isobutene may be copolymerized with 2 to 5% of isoprene to give butyl rubber, which was used for inner tubes before the advent of the tubeless tire because of its impermeability to air. It is still used for specialty truck tires, as an inner liner for tubeless tires, for truck inner tubes, for tire sidewall components, and for air cushions and bellows. Because it is saturated it resists aging and is therefore useful in constructions such as convertible tops. Low molecular weight polymers of isobutene are lubricating oil additives (Part II, Chap. 10).

Isobutene, like ethylene and propylene, may be hydrated to the corresponding alcohol, in this case the solid t-butyl alcohol, which is used on a small scale to prepare t-butylphenol for oil-soluble phenol-formaldehyde resins. Other uses are discussed in Sections 2.5.3 and 2.5.4.

$$CH_2{=}\underset{\underset{\displaystyle CH_3}{|}}{C}{-}CH_3 \xrightarrow{H_2O} CH_3{-}\underset{\underset{\displaystyle CH_3}{|}}{\overset{\overset{\displaystyle CH_3}{|}}{C}}{-}OH$$

Isobutene t-Butyl alcohol

Isobutene has allylic hydrogens like propylene and undergoes ammoxidation to methacrylonitrile. The possibility of oxidizing it to methacrylic acid has already been discussed (Sec. 2.5.3). It will alkylate p-cresol to give the important antioxidant, butylated hydroxytoluene (BHT).

p-Cresol   Isobutene   Butylated hydroxytoluene (BHT)

Liquid phase chlorination of isobutene leads to methallyl chloride. Treatment with aluminum and hydrogen gives triisobutyl aluminum and diisobutyl aluminum hydride, both of which are important in Ziegler-Natta polymerization catalysts (Sec. 4.3.10). The hydride is also a reducing agent.

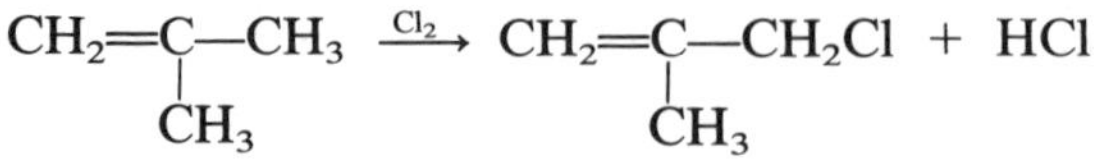

$$CH_2=\underset{\underset{CH_3}{|}}{C}-CH_3 \xrightarrow{Cl_2} CH_2=\underset{\underset{CH_3}{|}}{C}-CH_2Cl + HCl$$

Isobutene   Methallyl chloride

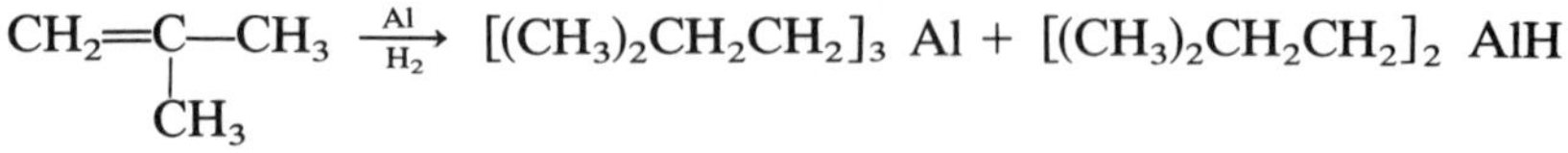

$$CH_2=\underset{\underset{CH_3}{|}}{C}-CH_3 \xrightarrow{Al}_{H_2} [(CH_3)_2CH_2CH_2]_3 Al + [(CH_3)_2CH_2CH_2]_2 AlH$$

Isobutene   Triisobutyl aluminum   Diisobutyl aluminum hydride

The only major tonnage use for isobutene is the declining, or at best static market, for butyl rubber (Part II, Sec. 4.2.2). In the United States any surplus can be and in fact is used for alkylation, but in Europe other outlets are being sought. One possibility, developed in France and said to be in use in the USSR, is the reaction of isobutene with formaldehyde to give dimethyldioxane, which can be cracked to isoprene (Part II, Sec. 4.1.1).

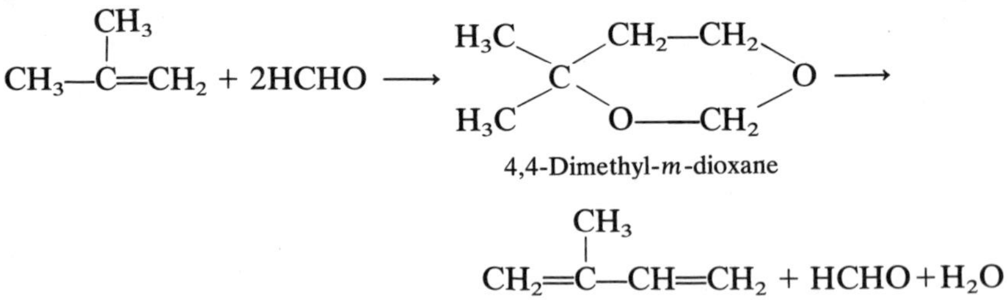

$$CH_3-\underset{\underset{CH_3}{|}}{C}=CH_2 + 2HCHO \longrightarrow$$

4,4-Dimethyl-*m*-dioxane

$$CH_2=\underset{\underset{CH_3}{|}}{C}-CH=CH_2 + HCHO + H_2O$$

Isoprene

In the West, isoprene is made by dehydrogenation of amylenes (2-methylbutene-1 and 2-methylbutene-2) extracted from the C_5 fraction obtained by catalytic cracking of crude oil to gasoline. A plant based on dimerization of propylene followed by isomerization and pyrolysis of the products was closed by Goodyear in 1978.

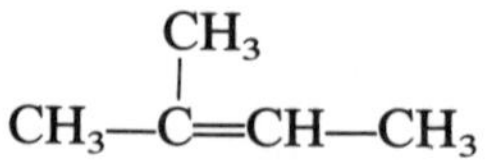

$$CH_3-\underset{\underset{CH_3}{|}}{C}=CH-CH_3$$

2-Methyl-2-butene

and

$$CH_2=\underset{\underset{CH_3}{|}}{C}-CH_2-CH_3 \longrightarrow CH_2=\underset{\underset{CH_3}{|}}{C}-CH=CH + H_2$$

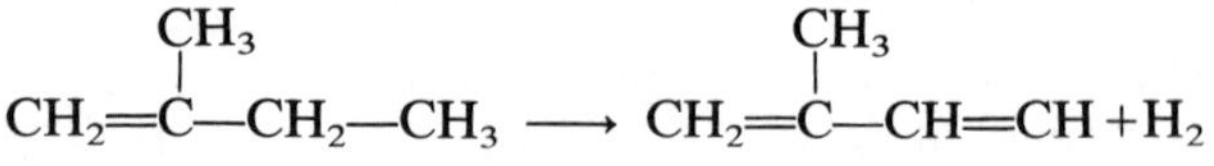

2-Methyl-1-butene

$$CH_3-CH{=}CH_2 \xrightarrow{\text{dimerization}} \underset{\text{2-Methyl-1-pentene}}{CH_2{=}\overset{\overset{\displaystyle CH_3}{|}}{C}-CH_2CH_2CH_3} \xrightarrow{\text{isomerization}}$$

$$\underset{\text{2-Methyl-2-pentene}}{CH_3-\overset{\overset{\displaystyle CH_3}{|}}{C}{=}CHCH_2CH_3} \xrightarrow{\text{Pyrolysis}} CH_2{=}\overset{\overset{\displaystyle CH_3}{|}}{C}-CH{=}CH_2 + CH_4$$

A process used in Italy starts with acetylene and acetone as follows:

$$HC{\equiv}CH + CH_3-\overset{\overset{\displaystyle CH_3}{|}}{C}{=}O \longrightarrow CH_3-\overset{\overset{\displaystyle CH_3}{|}}{\underset{\underset{\displaystyle OH}{|}}{C}}-C{\equiv}CH \xrightarrow{H_2}$$

2-Methyl-3-butynol-2

$$CH_3-\overset{\overset{\displaystyle CH_3}{|}}{\underset{\underset{\displaystyle OH}{|}}{C}}-CH{=}CH_2 \xrightarrow{-H_2O} CH_2{=}\overset{\overset{\displaystyle CH_3}{|}}{C}-CH{=}CH_2$$

2-Methyl-3-butenol-2

Like propylene and isobutene, 1-butene and 2-butene mixtures can be hydrated with concentrated sulfuric acid to give isobutanol or *sec*-butyl alcohol.

$$CH_2{=}CH-CH_2-CH_3 \xrightarrow{H_2O} CH_3-CH_2-\overset{\overset{}{\underset{\underset{\displaystyle OH}{|}}{CH}}}-CH_3$$
or

$$CH_3-CH{=}CH-CH_3$$
n-Butenes   *sec*-Butyl alcohol

$$\Big\downarrow {-H_2}$$

$$CH_3-CH_2-\overset{}{\underset{\underset{\displaystyle O}{\|}}{C}}-CH_3$$
Methyl ethyl ketone

Sec-butyl alcohol has a few minor uses as a solvent, but most of it is dehydrogenated to methyl ethyl ketone (MEK), an important solvent for vinyl and nitrocellulose lacquers and acrylic resins. In the United States a vapor phase process with a zinc oxide or zinc-copper catalyst is used, while in Europe a liquid phase reaction with finely divided nickel or copper chromite is preferred.

The mixture of 1- and 2-butenes can also undergo the oxo reaction to give a mixture of amyl alcohols, which are useful as solvents as such or as simple esters. As mentioned previously (Sec. 2.5.6) catalysts are available that would lead to a high percentage of the n-isomer with butene-1.

$$CH_2{=}CH{-}CH_2{-}CH_3 \xrightarrow[\text{catalyst}]{CO,\,H_2} \left[\begin{array}{l} CH_2{-}CH_2{-}CH_2{-}CH_3 \\ | \\ CHO \end{array} \right] +$$

Butene-1

$$\begin{array}{l} CH_3{-}CH{-}CH_2{-}CH_3 \\ \quad\quad | \\ \quad\quad CHO \end{array}$$

$$\downarrow H_2$$

$$\begin{array}{l} CH_3{-}CH{-}CH_2{-}CH_3 \\ \quad\quad | \\ \quad\quad CH_2OH \end{array}$$

Isoamyl alcohol

$$\downarrow H_2$$

$$CH_3{-}CH_2{-}CH_2{-}CH_2{-}CH_2OH$$

n-Amyl alcohol

$$CH_3{-}CH{=}CH{-}CH_3 \xrightarrow[\text{catalyst}]{CO,\,H_2} \left[\begin{array}{l} CH_3{-}CH{-}CH_2{-}CH_3 \\ \quad\quad | \\ \quad\quad CHO \end{array} \right]$$

Butene-2

$$\downarrow$$

$$\begin{array}{l} CH_3{-}CH{-}CH_2{-}CH_3 \\ \quad\quad | \\ \quad\quad CH_2OH \end{array}$$

Isoamyl alcohol

2-Butene can also undergo selective oxidation to maleic anhydride in the presence of molybdenum, vanadium, and phosphorus oxides (Fig. 2.18) as would be expected from the presence of two methyl groups with allylic hydrogens. The process is used in Japan but is reported to give poor yields. Maleic anhydride is more usually obtained by vapor phase oxidation of benzene over a vanadium oxide-molybdenum oxide catalyst. In principle the butene route should be the more economic because two carbon atoms are not lost as CO_2. The incentive of economy led to the consideration of n-butane as a feedstock for maleic anhydride, and vapor phase oxidation in the presence of phosphorus and vanadium salts does indeed provide maleic anhydride in good yield. Accordingly, at least two benzene-based plants have been converted to butane feedstocks, and it is expected that new capacity will use butane predominantly. It is impressive that a catalyst system can be found that will oxidize methyl groups to carboxyls and insert a double bond while refraining from breaking C—C bonds. Another process for maleic anhydride that has been considered uses as feed a mixture of n-butane and 1- and 2-butenes.

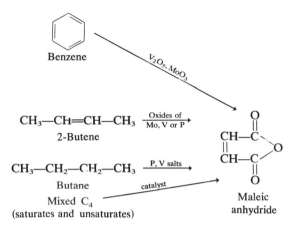

Figure 2.18 Maleic anhydride synthesis.

1-Butene and 2-butene may be separated by extractive distillation and polymerized separately. Polybutene-1 is useful in plastic, paper, and film applications.

C$_4$ hydrocarbon supplies will increase in the United States in the next decade as gas oil cracking becomes more prevalent. Indeed the C$_4$ fraction will probably shift from a relatively scarce commodity to one, which like propylene and toluene, is in surplus. Such surpluses provide a stimulus to industrial chemists, and much C$_4$ chemistry may develop. There are many leads in the literature such as the carbonylation of butadiene in the presence of methanol to provide an unsaturated precursor of adipic acid. The reaction takes place in the presence of a palladium-copper catalyst under mild conditions. Adipic acid is now made from benzene (Sec. 2.7.2), which is potentially in short supply because of its use in nonleaded gasoline. Here is a possible way of easing pressure on benzene.

$$CH_2{=}CH{-}CH{=}CH_2 \xrightarrow[\text{catalyst}]{CO,\ CH_3OH}$$
Butadiene

$$CH_3O{-}\underset{\underset{O}{\|}}{C}{-}CH_2{-}CH{=}CH{-}CH_2{-}\underset{\underset{O}{\|}}{C}{-}OCH_3 \xrightarrow[\text{hydrolysis}]{[H]}$$

$$HOOC{-}(CH_2)_4{-}COOH$$
Adipic acid

Furthermore, such syntheses involving carbon monoxide are attractive in the long term because this small molecule is potentially available from water gas, the building block most easily prepared from coal. Another interesting possibility, should the C_4 fraction become very cheap, is the dimerization of butadiene to vinylcyclohexene followed by dehydration to styrene.

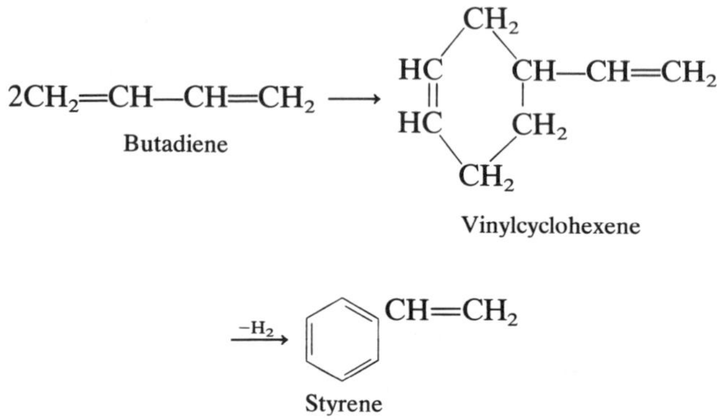

$$2CH_2{=}CH{-}CH{=}CH_2 \longrightarrow$$

Butadiene

Vinylcyclohexene

$$\xrightarrow{-H_2}$$ CH=CH_2

Styrene

2.7 CHEMICALS FROM BENZENE

In the United States most benzene comes from the catalytic reforming of light or heavy naphtha (Sec. 2.2.1) which yields a benzene-toluene-xylenes mixture. Benzene is also a volatile by-product of the conversion of coal to coke. Coal tar benzene is available only to the extent that coke is required by the steel industry, and this has not expanded to accommodate the expanding benzene market. Furthermore, coal tar benzene contains impurities such as thiophene that can be removed only with difficulty and that render it unacceptable for many purposes. In 1949 all US benzene was produced from coal tar; by 1959 the proportion had dropped to 50%. In 1972 it was down to only 6.4%, and it is now about 4.4%.

This is not the whole story however. Catalytic reforming typically leads to a BTX mixture containing about 50% toluene, about 35–45% xylenes, but only about 10–15% benzene, the end product most in demand. Production of necessary quantities of benzene would lead to an embarrassing surplus of toluene and xylenes. These could be returned to gasoline to raise its octane number, but this would mean that the total processing costs including separation of the three products would have to be borne by the benzene*. A process has been devised therefore to convert toluene to benzene by hydrodealkylation (Sec. 2.8), and this currently accounts for almost 30% of US benzene production.

The situation in Europe has been similar but displaced by a number of years. Petrochemical benzene did not appear in Britain until 1965 and accounted for about one-third of total European production at that time. The proportion is now about 90%. Also, steam cracking of naphtha, as practised in Europe and to a much lesser extent in the United States, gives a large amount of a gasoline fraction called pyrolysis gasoline, which contains a marketable balance of aromatics in which benzene predominates. Steam cracking of ethane, propane, and butane produces smaller amounts of pyrolysis gasoline, but this is still a source of benzene in both the United States and Europe. Pyrolysis gasoline is of much greater importance in Europe and accordingly hydrodealkylation is less significant there. In the United Kingdom it accounts for about 15% of the total.

In 1977, 11.25 billion lb of benzene was used as a chemical feedstock as compared to 12.6 billion lb of propylene. We have already mentioned some of its uses. The biggest is its reaction with ethylene to give ethylbenzene for styrene and for ethylbenzene hydroperoxide for propylene oxide and styrene preparation (Sec. 2.4.4). Next is its reaction with propylene to give cumene for conversion to phenol and acetone (Sec. 2.5.3). In 1977 the production of 7.3 billion lb of ethylbenzene and 2.64 billion lb of cumene accounted for over 65% of the total benzene consumption.

Cyclohexane, a feedstock for nylons, comes third after styrene and cumene as a consumer of benzene, but before discussing it we should say more about phenol. This is yet another chemical for which a variety of processes is available.

2.7.1 PHENOL

The earliest process for phenol was the sulfonation of benzene to the sulfonic acid followed by fusion of the acid with alkali (**a**, Fig. 2.19). During the Boer War (1899–1902) the British used picric acid (trinitrophenol), an uncertain and unreliable explosive, in their shells. Coal tar phenol supplies would not produce sufficient picric acid for the war effort, and the benzenesulfonate process thus became the first tonnage organic chemical process to be operated.

The second route appeared in 1924 (**b**, Fig. 2.19) and involved the

---

* Since this was written, petroleum companies are indeed adding pure toluene to gasoline to compensate for the removal of lead tetraethyl. This is reflected in the price of gasoline.

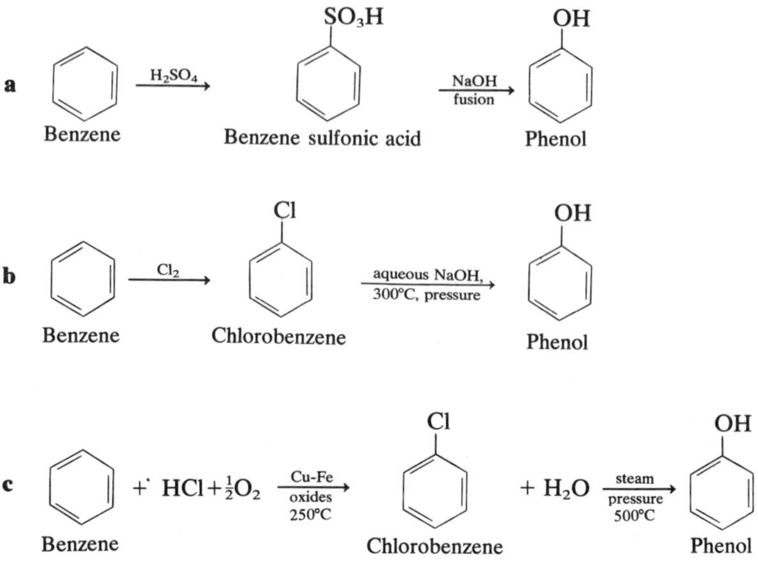

a Benzene → (H$_2$SO$_4$) Benzene sulfonic acid (SO$_3$H) → (NaOH fusion) Phenol (OH)

b Benzene → (Cl$_2$) Chlorobenzene (Cl) → (aqueous NaOH, 300°C, pressure) Phenol (OH)

c Benzene + HCl + $\frac{1}{2}$O$_2$ → (Cu-Fe oxides 250°C) Chlorobenzene (Cl) + H$_2$O → (steam pressure 500°C) Phenol (OH)

d Cumene hydroperoxide method see **h**, Fig. 2.10 and Fig. 2.12

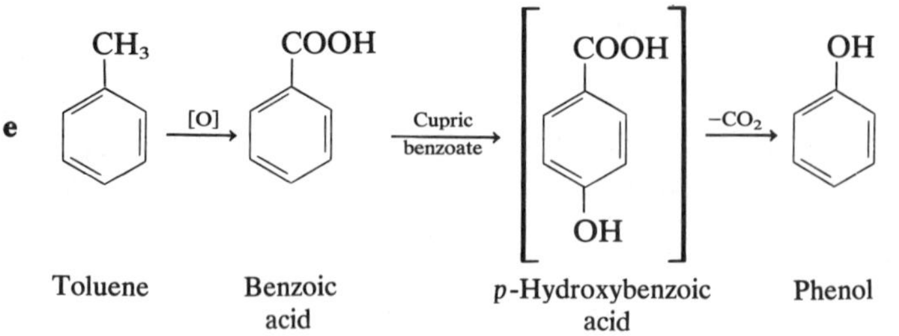

e Toluene (CH$_3$) → [O] Benzoic acid (COOH) → (Cupric benzoate) [p-Hydroxybenzoic acid (COOH ... OH)] → (−CO$_2$) Phenol (OH)

Figure 2.19   Phenol—methods of preparation.

direct chlorination of benzene to chlorobenzene, which was then hydro-lyzed to phenol by means of sodium carbonate or sodium hydroxide.

In a later process, the Raschig regenerative process (**c**, Fig. 2.19), chlorobenzene was prepared from HCl, air, and benzene (compare the oxychlorination of ethylene, Sec. 2.4.1) and hydrolyzed by steam in the vapor phase.

The cumene-phenol process (**d**, Fig. 2.19) dates from the early 1950s and has already been described (Sec. 2.5.3). The newest route is motivated by a surplus of toluene (**e**, Fig. 2.19) which is first oxidized in the liquid phase to benzoic acid. Molten benzoic acid then reacts with air and steam in the presence of cupric benzoate promoted with magnesium benzoate as catalyst. The volatile phenol is removed by distillation. The mechanism may involve intermediate formation of p-hydroxybenzoic acid, which decarboxylates to phenol. Other mechanisms have been proposed.

To be economical a benzenesulfonate plant requires cheap sulfuric acid and caustic soda and a nearby paper mill that can use the sodium sulfite by-product. Labor costs are high but capital costs are low. About 5% of US synthetic phenol is made in this way.

The chlorobenzene process is expensive because of chlorine and alkali usage and must be operated on a very large scale to be economical. The recent closing of Dow-Midland's massive 100,000 tons per year plant means that only about 3% of US synthetic phenol is now made in this way.

The Raschig process is not as wasteful of chlorine and alkali as the chlorobenzene route, but conversions per pass are low, and therefore capital costs are high. Also the high temperatures and acid conditions cause corrosion problems, and the high pressures raise operating costs. The process has been obsolete since 1971.

The toluene oxidation process is simple, and there are no by-products. Its economics depend on how cheaply toluene could otherwise be converted to benzene (Sec. 2.8). There is only one US plant, and it manufactures about 2% of US synthetic phenol.

The cumene process has much in its favor. No expensive chlorine, NaOH, or H_2SO_4 is wasted; conditions are mild; and utility costs are low. The only drawback is that there must be a demand for both chemicals produced. Currently this is no problem, and the route accounts for almost 90% of US synthetic phenol production. In the United Kingdom virtually all phenol is made by cumene peroxidation. In continental Western Europe the proportion is slightly lower than in the United States.

A plant was built in Australia in which cyclohexanol/cyclohexanone mixtures were dehydrogenated to phenol but it was soon closed as uneconomical. There has been research on direct air oxidation of benzene to phenol but it has not yet come to fruition.

The biggest use of phenol, the manufacture of phenol-formaldehyde resins (Part II, Sec. 2.3.1), accounts for half of US phenol consumption and a little over a third of that in Western Europe. Caprolactam (Sec. 2.7.2) is the second important derivative.

Phenol may be condensed with acetone to give bisphenol A which reacts with epichlorohydrin to give epoxy resins (Part II, Sec. 2.3.5), with phosgene or diphenyl carbonate to give polycarbonates (Part II, Sec.

2.1.8), and with several complex sulfones to give polysulfone resins (Part II, Sec. 2.1.8). For epoxy resins a mixture of the *o*- and *p*-isomers is acceptable, but the other two polymers require the *p,p*-isomer.

Phenol   Acetone   *p,p*-Bisphenol A

o,p-Bisphenol A

Phenol may also be alkylated with propylene trimer (or various other olefins) to give *p*-nonylphenol for conversion to detergents either by sulfonation or ethoxylation. The detergents have poor biodegradability, but the ethoxylated products have good low foam properties and are still used in Europe. In addition, nonylphenol is a basis for antioxidants for rubbers and plastics. Other alkylphenols find application in phenolic resins (Part II, Sec. 2.3.1).

Phenol also finds a small use as a raw material for the important herbicide 2,4-dichlorophenoxyacetic acid.

2,4-Dichlorophenol   2,4-Dichlorophenoxyacetic acid

2.7.2 NYLON

Benzene is the source of the two most important nylons, nylon 6,6 and nylon 6. Nylon 6,6 is the polymer formed when adipic acid condenses with hexamethylene diamine (**a**, Fig. 2.20). Nylon 6 (**b**, Fig. 2.20) on the other hand, is the self-condensation product of caprolactam, which is the

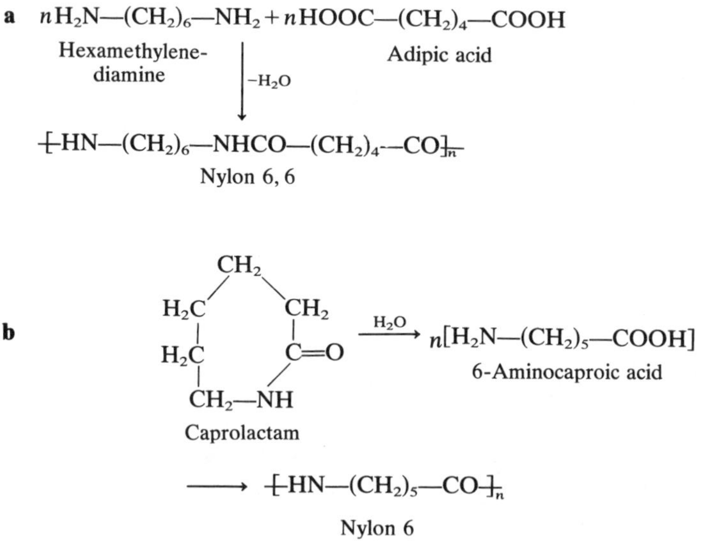

a $n \, H_2N—(CH_2)_6—NH_2 + n \, HOOC—(CH_2)_4—COOH$

Hexamethylene- diamine   Adipic acid

$\Big\downarrow \, {\scriptstyle -H_2O}$

$\{HN—(CH_2)_6—NHCO—(CH_2)_4—CO\}_n$

Nylon 6,6

b

$$\text{Caprolactam} \xrightarrow{H_2O} n[H_2N—(CH_2)_5—COOH]$$

6-Aminocaproic acid

Caprolactam

$$\longrightarrow \{HN—(CH_2)_5—CO\}_n$$

Nylon 6

Figure 2.20 Nylon structures.

dehydration product of 6-aminocaproic (6-aminohexanoic) acid. The numbers used to designate nylons refer to the number of carbon atoms in the diamine and dibasic acid in that order. A single number indicates that the amino and carboxyl functions are in one molecule.

In the early days of nylon DuPont created a mystique about it by pointing out that it was made from coal, air, and water. Coal at that time was the source of the benzene or phenol feedstocks. Now both are derived from petroleum and "made from petroleum, air, and water" does not sound nearly so good.

The first step in the manufacture of the adipic acid needed for nylon 6,6 (Fig. 2.21) is the hydrogenation of benzene to cyclohexane. Thereafter several routes are available. The cyclohexane could be oxidized directly to adipic acid with nitric acid or with air over cobalt acetate, but yields are low. Production of valueless by-products is high, and large amounts of nitric acid are consumed. Instead, a two-stage process is used. The initial oxidation over cobalt naphthenate or cobalt octoate gives a cyclohexanol/cyclohexanone "mixed oil." The proportion of cyclohexanol may be increased by the use of boric acid, which esterifies it as it is formed and prevents further oxidation to cyclohexanone. The cyclohexanol/cyclohexanone "mixed oil" may be oxidized to adipic acid with nitric acid and a catalyst system such as copper turnings with ammonium

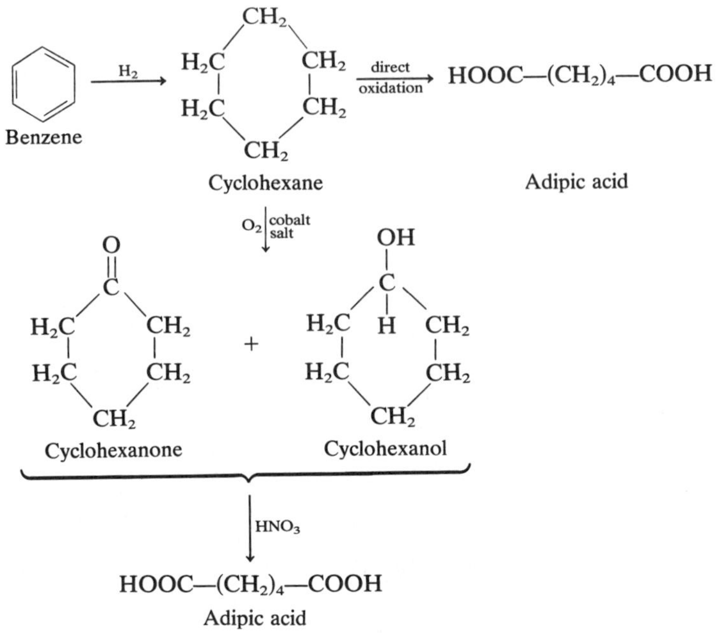

Figure 2.21 Adipic acid from benzene.

vanadate. In general, industrial oxidations are carried out with free air or cheap oxygen. This process is a notable exception, and no way of circumventing the use of nitric acid has yet been found.

Phenol may also be the starting point with catalytic hydrogenation yielding a mixture of cyclohexanol and cyclohexanone. Palladium-on-carbon gives the ketone, nickel the alcohol. As phenol is now more expensive than benzene the route is little used for nylon 6,6 but has advantages in nylon 6 production.

A route to adipic acid not currently used but drawing on an agricultural feedstock is based on tetrahydrofuran. The waste of chlorine and use of expensive and poisonous sodium cyanide suggest that even in an oil crisis it would not be particularly attractive. The tetrahydrofuran is derived from furfural (Sec. 3.3.1).

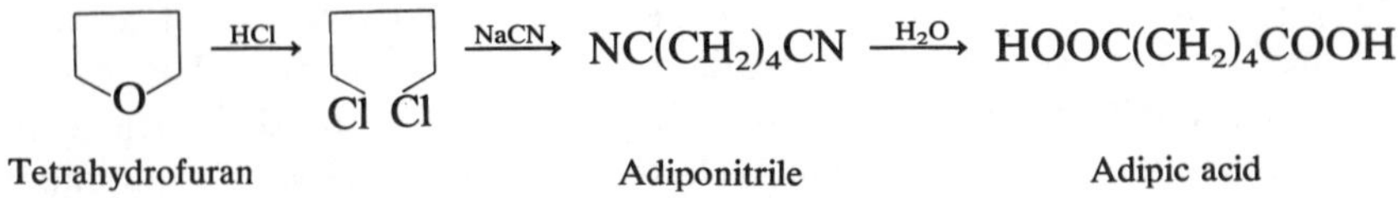

Once adipic acid has been made, the remaining monomer for nylon 6,6 is HMDA, and its manufacture has already been described (Sec. 2.6.2).

Nylon 6, like nylon 6,6, is manufactured from benzene, and there are again a number of competing routes. In the classic one the cyclo-hexanol/cyclohexanone "mixed oil" is dehydrogenated to cyclohexanone, which reacts with hydroxylamine to give cyclohexanone oxime (**a**, Fig. 2.22). Oximes undergo the Beckmann rearrangement, and this one is no exception; treatment with sulfuric acid gives caprolactam sulfate, which is converted by ammonia to caprolactam and ammonium sulfate.

The large amount of ammonium sulfate by-product, both from the final stage and from hydroxylamine production, soon saturated the fertilizer market (although this situation has recently changed) and stimulated the search for alternative methods. One of them involves the use of phosphoric rather than sulfuric acid to effect the Beckmann rearrangement because ammonium phosphate has greater value as a fertilizer.

The Japanese process shown in **b** of Figure 2.22 involves treatment of cyclohexane with nitrosyl chloride and hydrogen chloride under actinic light ($500\,m\mu$) to give the oxime hydrochloride directly. The major problem in the development of this synthesis was the engineering of an appropriate light source. This same reaction is useful on cyclododecane (Sec. 2.6.1) to provide the monomer for nylon 12.

Another process (**c**, Fig. 2.22) now largely of historical interest but used at one time by DuPont, involves nitration of cyclohexane to nitrocyclo-hexane. This can be reduced over a zinc-chromium catalyst to cyclo-hexanone oxime, the precursor of caprolactam. The manufacture of hydroxylamine and of half the ammonium sulfate is avoided. The nitration step is akin to the nitration of propane to nitroparaffins.

A fourth route to caprolactam (**d**, Fig. 2.22) uses peracetic acid (from acetaldehyde and oxygen) (Sec. 2.4.3) to convert cyclohexanone to capro-lactone, which on reaction with ammonia provides caprolactam.

Phenol is a workable raw material for cyclohexanone for nylon 6 because it can be hydrogenated in one stage, whereas benzene must go through cyclohexane and "mixed oil." Cyclohexane, furthermore, can be extracted from the naphtha fraction of petroleum by distillation prior to reforming to provide a third possible raw material. A toluene-based route to caprolactam has been pioneered in Italy. Benzoic acid (**e**, Fig. 2.22) prepared by the oxidation of toluene is hydrogenated to hexahydro-benzoic acid, and this is treated with nitrosylsulfuric acid, the reaction product of NH_3 absorbed in H_2SO_4. This may be the most economical of all the processes.

The international position is complicated. One US producer gets cyclo-hexane from the straight run gasoline or light naphtha fraction rather

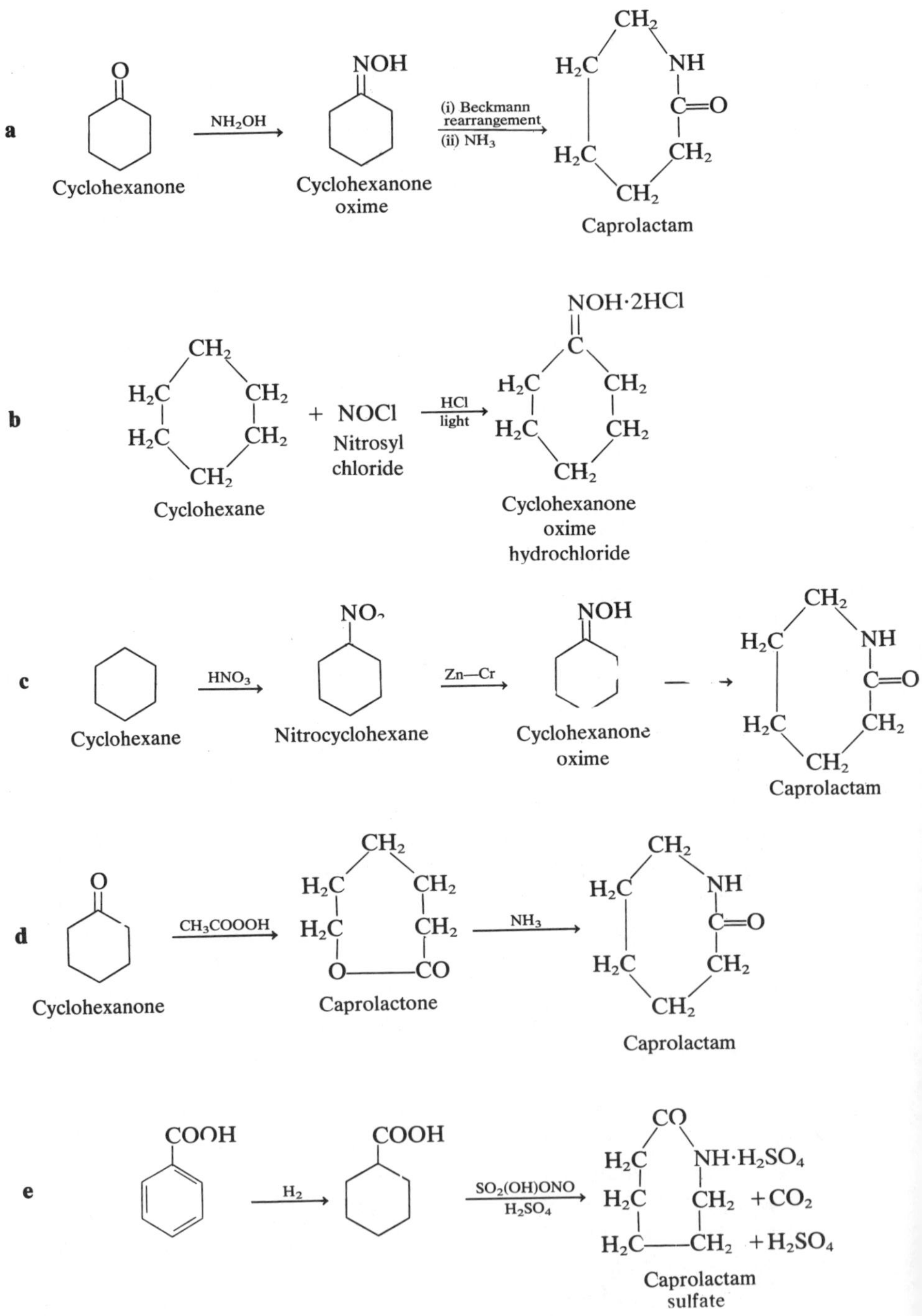

Figure 2.22 Possible routes to caprolactam.

than by benzene hydrogenation, but apparently no one else does. Very little adipic acid and hence nylon 6,6 is made from phenol anywhere in the world. In the United States benzene is slightly preferred to phenol for caprolactam manufacture, and on a world basis it accounts for about 75%.

In the United States and United Kingdom nylon 6,6 is the preferred polymer. In 1976 nylon 6 held 33% of the US market and 20% of the UK market, but in continental Western Europe it holds slightly over half the market with France and West Germany also contributing small quantities of nylons 11 and 12. In Japan nylon 6 holds 75% of the market. Small quantities of nylons 11 and 12 are also used (Sec. 3.2.8).

2.7.3   CHLOROBENZENE AND ANILINE

The most important nonpolymer use for benzene is for detergents, and these are discussed separately (Part II, Chap. 7).

Other uses for benzene are smaller. Chlorobenzene is an intermediate for phenol in an obsolescent process (Sec. 2.7.1) and condenses with chloral to give bis(chlorophenyl)trichloromethylmethane (DDT), an insecticide now banned in the United States.

$$2Cl-\bigcirc + \underset{\underset{Cl}{|}}{\overset{\overset{Cl}{|}}{Cl-C}}-CHO \longrightarrow Cl-\bigcirc-\underset{\underset{H}{|}}{\overset{\overset{Cl-C-Cl}{\overset{|}{Cl}}}{C}}-\bigcirc-Cl$$

Chlorobenzene   Trichloroacetaldehyde   Bis(chlorophenyl)
                                          trichloromethylmethane

Oxidation of benzene over vanadium pentoxide gives maleic anhydride (Sec. 2.6.3).

Nitration with "mixed acids" (H_2SO_4/HNO_3) gives nitrobenzene, which is reduced to aniline. The patent literature (see notes) describes a fascinating synthesis of aniline based on the direct reaction of benzene with ammonia. A so-called "cataloreactant," a term apparently coined for this particular reaction, comprises a catalytic amount of nickel and a stoichiometric amount of nickel oxide, which is reduced to nickel by the hydrogen that forms during the condensation. Rare earth metal oxides are used in very small quantities in the catalyst system.

$$\bigcirc + NH_3 \xrightarrow[\text{rare earth oxide}]{\text{Ni/NiO}} \bigcirc^{NH_2} + H_2$$

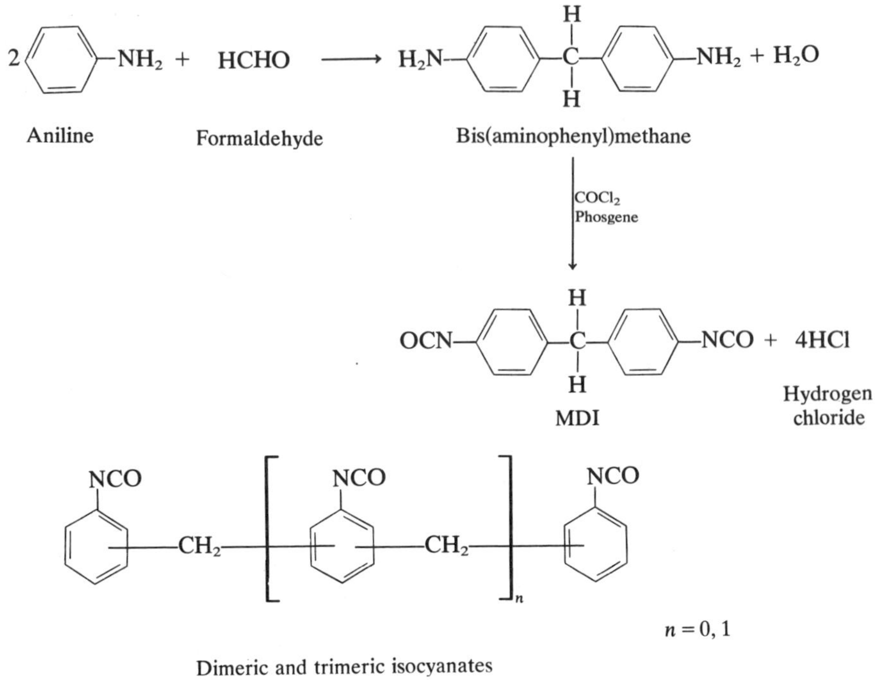

Figure 2.23 reaction scheme with Aniline, Formaldehyde, Bis(aminophenyl)methane, MDI, Hydrogen chloride, and Dimeric and trimeric isocyanates

$$2 \text{—NH}_2 + \text{HCHO} \longrightarrow \text{H}_2\text{N—} \overset{\text{H}}{\underset{\text{H}}{\text{C}}} \text{—NH}_2 + \text{H}_2\text{O}$$

Aniline   Formaldehyde   Bis(aminophenyl)methane

COCl₂
Phosgene

$$\text{OCN—} \overset{\text{H}}{\underset{\text{H}}{\text{C}}} \text{—NCO} + 4\text{HCl}$$

MDI   Hydrogen chloride

$n = 0, 1$

Dimeric and trimeric isocyanates

Figure 2.23   Dimeric and trimeric isocyanates.

Formerly the cornerstone of the dyestuff industry, aniline is now more widely used as an intermediate for rubber chemicals and, even more important, for isocyanates. The condensation with formaldehyde provides bis(aminophenyl)methane as well as dimers and trimers, which are represented in Figure 2.23. Reaction of these compounds with phosgene provides the corresponding isocyanates known as methylene diphenyl isocyanates (MDI). Dimers and trimers of MDI are used much more widely than the monomer, the biggest use being for rigid foams (Part II, Sec. 2.7).

2.8   CHEMICALS FROM TOLUENE

Toluene, like benzene, was once obtained from coal and is now obtained mainly from petroleum. It occurs in embarrassingly large quantities in the BTX fraction as a product of catalytic reforming as practised in the United States and in more manageable quantities in the pyrolysis gasoline fraction produced in European steam-naphtha cracking (Sec. 2.2.1).

With a US production in 1977 of 7.7 billion lb for chemical uses, toluene can claim to be the fourth most important petrochemical after ethylene, benzene, and propylene. An additional 51 billion lb (more than six times as much) produced largely in catalytic reforming was not isolated but was returned to the gasoline pool.

Because toluene, like propylene, is readily and cheaply available, chemists have worked to find uses for it. Of particular importance is its hydrodealkylation to benzene. Its disproportionation to benzene and mixed xylenes is carried out on a smaller scale. Its conversion to phenol (Sec. 2.7.1) and to caprolactam (Sec. 2.7.2) are further outlets. Its conversion to styrene and to terephthalic acid is described below.

Well over half of the toluene isolated in the United States is hydrodealkylated to benzene. This reaction may be purely thermal or it may be catalyzed by metals or supported metal oxides. Typical reaction conditions are 550–620°C over a chromium-alumina or platinum-alumina catalyst in the presence of hydrogen. The importance of hydrodealkylation depends on the relative prices of benzene and toluene at any given time.

$$\text{C}_6\text{H}_5\text{CH}_3 + \text{H}_2 \longrightarrow \text{C}_6\text{H}_6 + \text{CH}_4$$

A variation of hydrodealkylation is disproportionation, shown in **a** of Fig. 2.24. Disproportionation is important as a source of mixed xylenes from which the para isomer, the one most in demand, can be isolated. Two molecules of toluene react to give one of benzene and one of mixed xylenes. The reaction is carried out in the vapor phase in the presence of a nonnoble metal catalyst. Even a casual estimate of the energetics of the reaction indicates that the equilibrium constant is close to unity, so large volumes of reactants must be recycled from the product-recovery section of the plant to the reactor. The advantage of the process, however, is that the reactor effluent is free from ethylbenzene, which makes the separation of the xylene isomers easier. Although the reaction is cumbersome the demand for p-xylene has made it economical.

A further disproportionation reaction devised to obtain more p-xylene is called transalkylation (**b**, Fig. 2.24). Toluene is allowed to react with trimethylbenzene, which is formed during catalytic reforming. A methyl group migrates from a trimethylbenzene molecule to a toluene molecule giving two molecules of mixed xylenes. In practice benzene is also produced, but its volume can be kept low by use of a high ratio of trimethylbenzene to toluene. One US company operates these processes

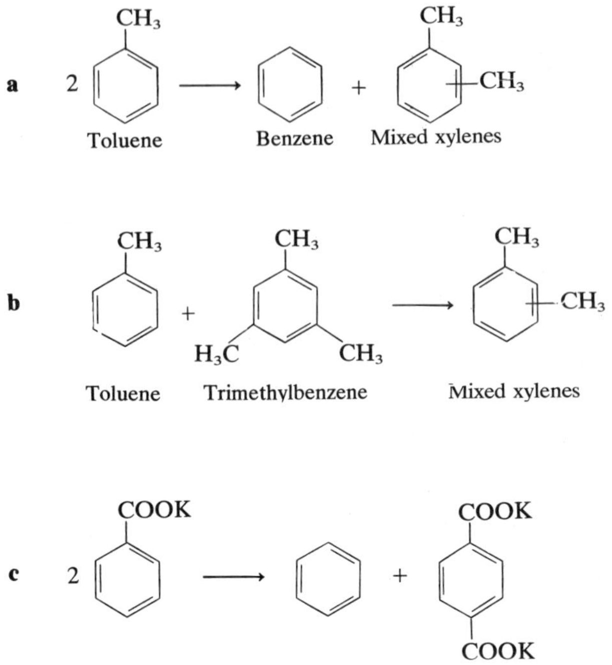

Figure 2.24 Disproportionation of toluene.

although their economics do not seem favorable unless toluene cost is low and the p-xylene price high.

A more interesting disproportionation, which gives only the p-isomer, starts with potassium benzoate (**c**, Fig. 2.24) obtained by oxidation of toluene to benzoic acid and neutralization of the latter. Potassium benzoate disproportionates to benzene and potassium terephthalate under a variety of conditions including temperatures over 400°C and modest pressures of 10–15 atm of CO_2. Catalysts include cadmium benzoate. The process is currently being commercialized. A problem that has prevented commercialization in the past but which apparently has been solved involves conversion of the salt to the free acid in such a way that the expensive potassium can be recycled. Acidification in the normal way yields a potassium salt of little value.

A further reaction developed partly in response to the toluene surplus is its conversion to styrene. This third method for making styrene (Sec. 2.4.4) has not yet been commercialized. It depends on the dehydro-coupling of toluene to stilbene followed by a metathesis (Sec. 2.5.7) reaction with ethylene. The first step takes place at 600°C in the presence of a lead magnesium aluminate catalyst and the second at 500°C with a calcium oxide-tungsten oxide catalyst on silica.

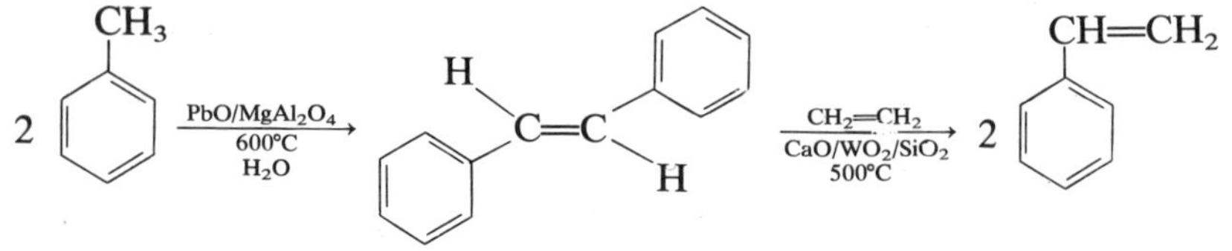

Stilbene

Styrene

Most toluene is consumed in processes devised to take up surplus production. A few depend on its specific properties. Toluene's second largest use is as a solvent largely for coatings. A federal decision in the United States that benzene is potentially carcinogenic will no doubt increase toluene's solvent usage at the expensive of benzene's, although the latter is small. Another application is as a raw material for a mixture of 2,4 and 2,6-tolylene diisocyanate (toluene diisocyanate, TDI) used for polyurethane resins (Part II, Sec. 2.7). TDI is made by dinitration of toluene followed by reduction of the nitro to amino groups. These react with phosgene to give the diisocyanate.

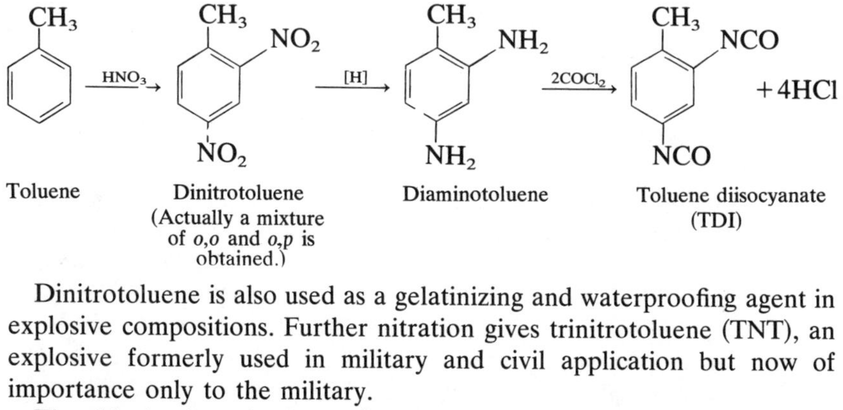

Toluene

Dinitrotoluene
(Actually a mixture
of *o,o* and *o,p* is
obtained.)

Diaminotoluene

Toluene diisocyanate
(TDI)

Dinitrotoluene is also used as a gelatinizing and waterproofing agent in explosive compositions. Further nitration gives trinitrotoluene (TNT), an explosive formerly used in military and civil application but now of importance only to the military.

The chlorination of toluene (Fig. 2.25) leads to benzyl chloride, which may be used to quaternize tertiary amines such as lauryl dimethylamine to give compounds used as germicides. Its main use is as a raw material for the PVC plasticizer, butyl benzyl phthalate. Dichlorination of toluene leads to benzal chloride, which is an intermediate for benzaldehyde, an ingredient of flavors and perfumes. Benzaldehyde may also be made by direct oxidation of toluene.

Chlorotoluenes are also intermediates in dyestuff manufacture, while sulfochlorination of toluene to toluenechlorosulfonic acid provides a

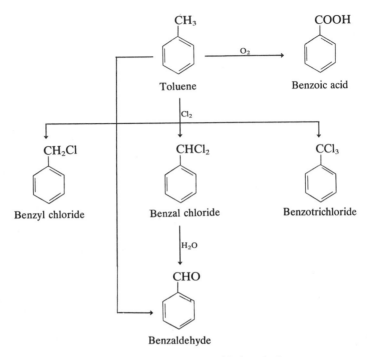

Figure 2.25 Chlorination and oxidation of toluene.

route to the corresponding sulfonamide, the starting material for saccharin.

Oxidation of toluene gives benzoic acid. This is a halfway stage in one process for phenol manufacture (Sec. 2.7.1), but benzoic acid itself has a few small uses. Diethylene glycol dibenzoate is a useful nonstaining plasticizer for PVC flooring where it competes with butyl benzyl phthalate. Benzyl benzoate is a perfume ingredient. Benzoic acid is an intermediate in a caprolactam synthesis (Sec. 2.7.2) and in a new process for production of terephthalic acid (**c**, Fig. 2.24)

2.9 CHEMICALS FROM XYLENES

Xylenes are produced mainly by catalytic reforming (Sec. 2.2.1). Insignificant quantities (about 1–2% of total production) are extracted from coal tar distillate. In Europe appreciable quantities are isolated from the pyrolysis gasoline that results from steam cracking of naphtha.

The separation of xylenes is outlined in Section 2.2.1. Other processes include selective absorption and clathration. p-Xylene is the isomer most in demand but is usually the smallest volume isomer in reformate. Isomerization, (Sec. 2.2.5.) disproportionation, and transalkylation reactions (Fig. 2.24) help to remedy this situation.

Xylenes are used as solvents, particularly in the paint industry, and any surpluses are returned to the gasoline pool.

Oxidation is the major chemical fate of the xylenes. p-Xylene gives terephthalic acid; m-xylene gives isophthalic acid; and o-xylene gives phthalic anhydride.

o-Xylene   Phthalic anhydride

m-Xylene   Isophthalic acid

p-Xylene   Terephthalic acid

Phthalic anhydride may also be obtained by oxidation of petrochemical or coal tar naphthalene over a vanadium pentoxide catalyst, a reaction analogous to the oxidation of benzene to maleic anhydride (Sec. 2.6.3). The loss of two carbon atoms as carbon dioxide seems to discriminate against this process, but yields from xylene oxidation are relatively low, and the naphthalene route appears to be holding its own. It is responsible for about one-third of US production and one-quarter of West European and Japanese production.

$+ 4\frac{1}{2}O_2 \xrightarrow{V_2O_5}$ $+ 2CO_2$

Naphthalene   Phthalic anhydride

About half of phthalic anhydride production is used in plasticizers (Part II, Chap. 11). One-quarter goes into alkyd resins (Part II, Sec. 5.6) and an equal amount into unsaturated polyesters (Part II, Sec. 2.3.4). Usage in alkyds is likely to remain steady; usage in polyesters will probably increase.

Isophthalic acid is not used on a large scale, and most *m*-xylene is returned to the gasoline pool. Isophthalic acid's main uses are in alkyd resins and unsaturated polyester resins where it provides improved solvent and alkali resistance. It reacts with *m*-phenylene diamine to give a specialty polymer, Nomex, which has high-temperature-resistant properties (Part II, Sec. 2.9.1).

Terephthalic acid is not only the raw material for polyester fibers such as Dacron or Terylene but also for a molding resin, produced in much smaller quantities, poly(butylene terephthalate). It results from the condensation of terephthalic acid with 1,4-butanediol.

2.10   METHANE DERIVATIVES

So far we have described six petroleum products as raw materials for chemicals—ethylene, propylene, the butylenes, benzene, toluene, and the xylenes. To complete the picture we must add methane from which we obtain hydrogen cyanide, chlorinated methanes, acetylene, and so-called synthesis gas, a mixture of hydrogen and carbon monoxide.

Acetylene and synthesis gas came originally from coal but methane is now the more important source and accordingly we shall describe them here.

The major source of methane in the United States and the United Kingdom is natural gas. It is also available from refinery gases but is recovered only in large refineries. US regulations fix the price at which natural gas may be shipped across state boundaries. This low price is said to have inhibited natural gas production in the US, although there is no doubt that reserves are diminishing.

2.10.1   HYDROGEN CYANIDE

Hydrogen cyanide was formerly made by the reaction of sodium, carbon (in the form of charcoal), and ammonia. In the first stage of the reaction ammonia and sodium reacted to form sodamide, $NaNH_2$. This then reacted with carbon to form sodium cyanamide, Na_2CN_2, which in turn reacted with more carbon to provide sodium cyanide, $NaCN$. Temperatures as high as 850°C were used, and yields were high based on the sodium and ammonia consumed. Reaction of the sodium cyanide with sulfuric acid provided hydrogen cyanide. The process has been superseded by the more economical reaction between methane, air, and ammonia over a platinum-rhodium catalyst at 1000°C. In a modified process air is omitted, and methane (or any other hydrocarbon feedstock up to naphtha) reacts directly with ammonia at 1400°C.

$$2Na + 2C + 2NH_3 \longrightarrow 2NaCN + 3H_2O \xrightarrow{H_2SO_4} HCN$$

Sodium cyanide   Hydrogen cyanide

$$2CH_4 + 3O_2 + 2NH_3 \xrightarrow[1000°C]{Pt-Rh} 2HCN + 6H_2O$$

Methane   Ammonia   Hydrogen cyanide

Hydrogen cyanide is also a by-product of the ammoxidation of propylene to acrylonitrile (Sec. 2.5.2). Because production is tied to that of acrylonitrile it must be disposed of somehow and has therefore taken over half of the market from HCN produced by the above processes.

Hydrogen cyanide and its sodium salt undergo a number of industrially important reactions. The use of HCN in hexamethylene diamine production is described in Section 2.6.2. This is its second largest use. Its largest use is for methyl methacrylate (Sec. 2.5.3). It has been used in two obsolete routes to acrylic acid (Sec. 2.5.1).

Sodium nitrilotriacetate was suggested a few years ago to replace the sodium tripolyphosphate "builder" in detergents (Part II, Sec. 7.6.9). It is made from ammonia, methanol, and HCN. It is manufactured in the United States for use as a chelating agent and for export but is not used in US detergent formulations because of possible toxicity problems.

$$NH_3 + 3CH_3OH + 3HCN \longrightarrow N(CH_2CN)_3 + 3H_2O + 3H_2$$

$$N(CH_2-CN)_3 \xrightarrow[H_2O]{H^+} N(CH_2-COOH)_3 + 3NH_4^+$$

Other uses for HCN include preparation of ethylenediamine tetraacetic acid and conversion to NaCN. A new application is the production of cyanuric chloride. This chemical is a starting material for the important triazine herbicides and fiber-reactive dyes. Cyanuric chloride is also useful for the manufacture of triallylcyanurate useful for polyester manufacture and in certain pharmaceutical syntheses.

$$HCN + Cl_2 \longrightarrow N≡C-Cl + HCl \qquad 3N≡C-Cl \longrightarrow$$

Cyanogen chloride

Cyanuric chloride

2.10.2 CHLORINATED METHANES

Chlorination of methane yields a mixture of mono, di, tri, and tetra-chlorides. Methyl chloride is chlorinated more rapidly than methane itself, so if methyl chloride is the desired product a ratio of methane to chlorine of at least 10:1 is required. Consequently the preferred route to methyl chloride is the reaction of methanol with HCl. This contrasts with ethane which can be chlorinated to ethyl chloride without difficulty, the second step to dichloroethane being slower than the first. Hydro-chlorination of ethylene (I, Fig. 2.5) is however the preferred route to ethyl chloride.

$$CH_4 \xrightarrow{Cl_2} CH_3Cl \xrightarrow{Cl_2} CH_2Cl_2 \xrightarrow{Cl_2} CHCl_3 \xrightarrow{Cl_2} CCl_4$$

| Methyl chloride +HCl | Methylene chloride +HCl | Chloroform +HCl | Carbon tetrachloride +HCl |

$$CH_3OH + HCl \longrightarrow CH_3Cl + H_2O$$
Methyl chloride

The more highly chlorinated methanes—methylene chloride, chloroform, and carbon tetrachloride—may be separated by distillation. A competitive route to carbon tetrachloride involves chlorination of carbon disulfide.

$$CS_2 + 3Cl_2 \longrightarrow CCl_4 + S_2Cl_2$$
Carbon disulfide   Carbon tetrachloride

$$2S_2Cl_2 + CS_2 \longrightarrow CCl_4 + 6S$$

Methyl chloride is used in the production of tetramethyl lead, silicone resins, and quaternary ammonium compounds (Sec. 3.2.2). Methylene chloride is mainly used as a paint stripper, especially for jet aircraft which must be stripped and examined for cracks annually (Part II, Sec. 9.5). Methylene chloride has the great advantage over alkaline paint removers of not attacking aluminum. On the other hand, many chlorinated hydrocarbons are under suspicion as health hazards. CCl_4 inhalation, for example, causes cirrhosis of the liver.

Chloroform, once an important anesthetic, is now used for production of fluorocarbon refrigerants, aerosol propellants, and fluorocarbon polymers. In tonnage it is the most important of the chlorinated methanes. The two main fluorocarbons are dichlorofluoromethane and trichlorofluoromethane, but fluorocarbons too are in disrepute, at least for use in aerosols, because of long-term effects they may have on the ozone layer (Sec. 6.4.4).

2.11 ACETYLENE

Acetylene rose in importance as a chemical feedstock after World War II, reached a peak in the mid-1960s, and has since declined. Coal was not only a source of aromatics, by way of coal tar, but was also the raw material for acetylene and hence the source of many early plastics and aliphatic organic chemicals.

Coke and lime are heated in an electric furnace to 2000°C to give calcium carbide. This is hydrolyzed by water to give acetylene. The furnace requires a huge input of electrical energy, and it has many of the characteristics of a nineteenth century industry. It is a batch process; it is labor intensive because of the handling of solids required; and it is environmentally unattractive because every ton of acetylene produced is accompanied by 2.8 tons of calcium hydroxide, usually in a slurry with 10 times as much water. The capital costs of the furnace are high and it has a short life because of the extreme conditions used.

$$CaO + 3C \longrightarrow \underset{\substack{\text{Calcium} \\ \text{carbide}}}{CaC_2} + CO$$

$$CaC_2 + 2H_2O \longrightarrow \underset{\text{Acetylene}}{CH{\equiv}CH} + Ca(OH)_2$$

Even in the days of cheap energy these factors provided an impetus for the replacement of the calcium carbide route. In the early 1960s, many attempts were made to derive acetylene from petroleum. Thermodynamically it is the most stable hydrocarbon above 1300°C, but the pyrolysis of other hydrocarbons is not so simple. It is difficult to quench the reaction mixture rapidly enough to prevent reverse reactions and to keep residence times short enough to prevent coking. Separation of acetylene from by-product hydrogen is a further difficulty.

These problems were addressed in the Wulff and Sachsse processes, which in theory can start with naphtha or natural gas. The naphtha-based plants, mainly European, gave continuous trouble and were closed after short lives, with the exception of a Northern Ireland plant in operation in 1977. On the other hand, natural gas cracking accounted for 54% of US acetylene production in 1973. Ninety-five percent of US coal-based acetylene is used in nonchemical applications such as arc welding, and the acetylene still used as a chemical feedstock is derived almost entirely from natural gas.

$$2CH_4 \; \rightleftharpoons \; CH{\equiv}CH + 3H_2$$
$$\text{Acetylene}$$

$$C_nH_{2n+2} \; \rightleftharpoons \; \tfrac{n}{2}CH{\equiv}CH + (\tfrac{n}{2}+1) \; H_2$$

Calcium carbide has played an honorable role in the world chemical industry and even gave its name to the Union Carbide Corporation, one of the largest US chemical companies. Its derivative, calcium cyanamide, gave its name to the Cyanamid Corporation. Its energy requirements mean, however, it can only be made in countries like Norway, Switzerland, and Japan with cheap hydroelectricity.

Even the Wulff and Sachsse processes are energy intensive and require temperatures of 1200–1600°C. Thus acetylene has become steadily more expensive in comparison with ethylene, and almost everything that used to be done with acetylene can now be done more cheaply with ethylene.

The list of obsolete acetylene-based processes is an impressive one—vinyl chloride from $C_2H_2 + HCl$, vinyl acetate from $C_2H_2 + CH_3COOH$, acrylonitrile from $C_2H_2 + HCN$, and acetaldehyde from $C_2H_2 + H_2O$. Chloroprene from acetylene dimer (vinylacetylene), acrylate esters by way of $C_2H_2 + ROH + CO$, and perchloroethylene by a multistage process are obsolescent. Between 1967 and 1974, 25 acetylene plants in the United States closed, and consumption for chemical use dropped from 1.23 to 0.40 billion lb between 1962 and 1977.

What acetylene chemicals are still important? The route to acrylates is still in use (Sec. 2.5.1). Trichloroethylene (for degreasing) and perchloroethylene (for drycleaning) may be made from acetylene. In the United States however, most perchloroethylene is produced from simultaneous chlorination and pyrolysis of hydrocarbons or by oxychlorination of dichloroethylene.

$$HC{\equiv}CH \xrightarrow{Cl_2} HCCl_2{-}CHCl_2 \xrightarrow{-HCl}$$
$$\text{Tetrachloroethane}$$

$$CCl_2{=}CHCl \xrightarrow{Cl_2} CCl_3{-}CHCl_2 \xrightarrow{-HCl} CCl_2{=}CCl_2$$
$$\text{Trichloroethylene} \qquad \text{Pentachloroethane} \qquad \text{Perchloroethylene}$$

A reaction of interest involves addition of hydrogen fluoride to acetylene to give vinyl fluoride that can be polymerized to polyvinyl fluoride, a polymer with outstanding weathering properties.

The major area where acetylene is still significant is in Reppe chemistry. Reppe was a German chemist who studied at the time of World War

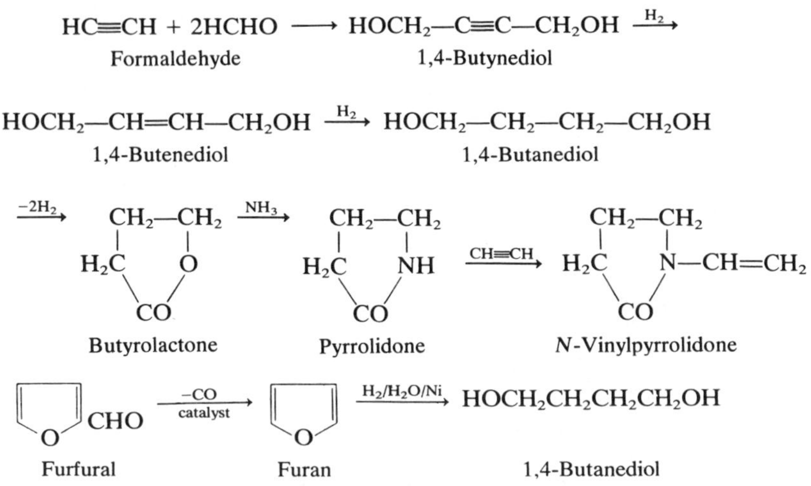

$$HC\equiv CH + 2HCHO \longrightarrow HOCH_2-C\equiv C-CH_2OH \xrightarrow{H_2}$$

Formaldehyde   1,4-Butynediol

$$HOCH_2-CH=CH-CH_2OH \xrightarrow{H_2} HOCH_2-CH_2-CH_2-CH_2OH$$

1,4-Butenediol   1,4-Butanediol

Butyrolactone   Pyrrolidone   N-Vinylpyrrolidone

Furfural   Furan   1,4-Butanediol

Figure 2.26 N-Vinylpyrrolidone synthesis.

II the interaction of acetylene under high pressure with aldehydes, ketones, alcohols, and carbon dioxide. The acrylate process mentioned above is a Reppe reaction.

Typical also of Reppe chemistry is the reaction of acetylene with formaldehyde to give 1,4-butynediol, which may be hydrogenated stepwise to butenediol and butanediol (Fig. 2.26). The latter compound is of particular interest in polyesters and may also be made from furfural (Fig. 2.26). Dehydrogenation of 1,4-butanediol gives butyrolactone, which reacts with ammonia to give pyrrolidone. Treatment with acetylene gives N-vinylpyrrolidone, which may be polymerized to polyvinylpyrrolidone. The latter is a water-soluble polymer that has a cosmetics application in hair sprays. During World War II it was of interest as a possible replacement for blood plasma.

2.12   SYNTHESIS GAS

The term "synthesis gas" refers to various mixtures of carbon monoxide and hydrogen used for the manufacture of chemicals. In the past it was produced by passage of steam over coke to give water gas (Sec. 3.1.2 and note to Sec. 3.1) and the composition was then adjusted for the desired synthesis. It has always been important in the chemical economy, and because it can be made from coal its importance could grow if petroleum shortages become acute. Many important chemicals can be made from it.

Petrochemical synthesis gas is usually derived from methane or naphtha but could be made just as easily from ethane, propane, or butane. The process is known as steam reforming and involves the passage of the hydrocarbons plus steam over a nickel catalyst. The desirable reaction, if methane is the feedstock, is

$$CH_4 + H_2O \rightleftharpoons CO + 3H_2$$

The undesirable reactions are

$$CO + H_2O \rightleftharpoons CO_2 + H_2 \quad \text{(water gas shift reaction)}$$

$$2CO \rightleftharpoons C + CO_2$$

$$CO + H_2 \rightleftharpoons C + H_2O$$

$$CH_4 \rightleftharpoons C + 2H_2$$

Carbon dioxide and carbon are both a nuisance, the latter particularly so because it deactivates the catalyst.

2.12.1 AMMONIA AND ITS DERIVATIVES

The most important use of synthesis gas is in the Haber process for the preparation of ammonia, a process that requires a $N_2 : 3H_2$ mixture but no carbon monoxide. The synthesis gas is mixed with air, which provides the nitrogen for the synthesis, and the carbon monoxide is converted to dioxide by the water gas shift reaction. The CO_2 is dissolved in water under pressure and the solution removed. The $N_2 : 3H_2$ mixture is passed at 450°C and 250 atm over a promoted iron oxide catalyst:

$$3H_2 + N_2 \rightleftharpoons 2NH_3$$

The Haber process was developed in Germany just before World War I in response to forecasts of a shortage of "fixed" nitrogen. It proved itself during the war, when German supplies of Chilean nitrates for fertilizers and explosives were cut off by the allied blockade. Detailed discussion of it would be out of place in a book on the organic chemicals industry, but some comment on the role of ammonia in the industry is necessary.

Ammonia and its inorganic salts are the most important nitrogenous fertilizers. Ammonia reacts with carbon dioxide to give ammonium carbamate, which in turn dehydrates to urea.

$$NH_3 + CO_2 \longrightarrow \underset{\substack{\text{Ammonium} \\ \text{carbamate}}}{NH_2COONH_4} \xrightarrow{-H_2O} \underset{\text{Urea}}{NH_2CONH_2}$$

Urea contains 46% fixed nitrogen and is a fertilizer in its own right particularly in the United States where 75% of the production is used for this purpose. Urea reacts with formaldehyde to give urea-formaldehyde (U/F) thermosetting resins. The structure of the thermoset urea-formaldehyde resin is complex, involving first of all the formation of methylolurea and N,N'-dimethylolurea.

$$H_2NCONHCH_2OH \qquad OC(HNCH_2OH)_2$$

Methylolurea   N,N'- Dimethylolurea

These, by a series of condensations that include ring formation, provide complex thermoset polymers whose structures are poorly defined (Sec. 4.4.1).

Urea is also the source of melamine, for all melamine in the United States is made by a one-step process in which molten urea is trimerized with ammonia in the presence of an aluminosilicate catalyst. Both low and high pressure processes are used, and in the low pressure process cyanic acid, HNCO, is an intermediate.

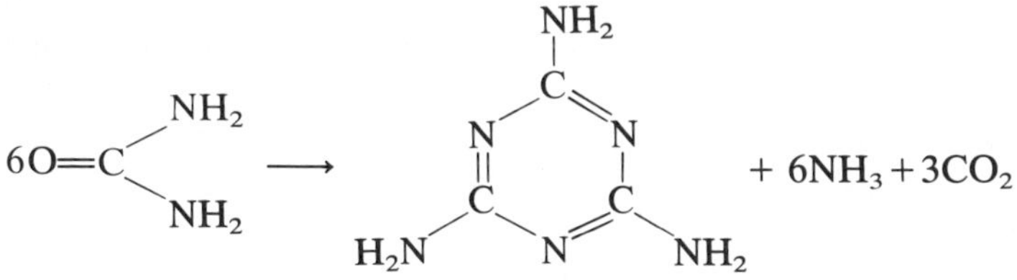

Melamine

An obsolete process started with calcium carbide from coal and went by way of dicyandiamide:

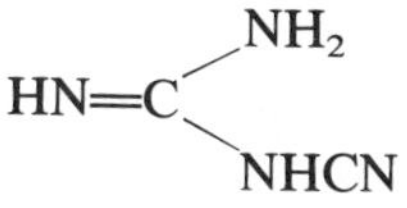

The rest of the world is moving toward the one-step process although two-stage plants may be in operation for some time yet.

Like urea, melamine reacts with formaldehyde to give complex thermosetting resins. M/F resins are highly quality premium products and are used for dinnerware, the top layer of laminates such as "formica," and in industrial coatings. Both U/F and M/F resins are useful for textile and paper treatment, adhesives, and molding powders.

Ammonia may be oxidized to nitric acid, which is used (sometimes mixed with sulfuric acid) to produce a variety of nitro compounds and their derivatives. Most explosives [e.g., cellulose nitrate, TNT, "Tetryl," picric acid, pentaerythritol tetranitrate, "Cyclonite" (cyclotrimethylene trinitramine), nitroglycerine, and ammonium nitrate] are nitro compounds as are nitromethane, nitroethane, and the nitropropanes, which are used as propellants, chemical intermediates, and solvents. Nitrobenzene is the precursor of aniline, and dinitrotoluene the precursor of tolylene diisocyanate (Sec. 2.8). Nitrocyclohexane was an intermediate in a caprolactam process (Sec. 2.7.2).

Ammonia also participates in ammoxidation reactions (Sec. 2.5.2) and can be reacted with acids to give ammonium salts which dehydrate to amides and to nitriles which in turn may be reduced to amines. In Section 2.6.1 we described how adiponitrile, a precursor of HMDA, was originally made from adipic acid. A similar reaction can be carried out with fatty acids. Ammonia will also react with alkyl halides to give amines, but the route is not popular because of the waste of an expensive halogen atom.

2.12.2   METHANOL AND FORMALDEHYDE

The second important use for synthesis gas, after ammonia production, is methanol production. The plant is often integrated with an ammonia plant both because the processes and ancillary equipment are similar and because the methanol plant can use the CO_2 removed from synthesis gas in ammonia synthesis. It is reacted with methane and steam over a promoted nickel catalyst at 800°C.

$$3CH_4 + CO_2 + 2H_2O \longrightarrow 4CO + 8H_2$$

In the traditional process the $2H_2:CO$ mixture passed over a mixed oxide catalyst (e.g., zinc oxide plus 10% chromium oxide) at 300°C and 250 to 300 atm. New plants operate a lower pressure process developed in the United Kingdom. A high volume reactor, a pressure between 50 and 100 atm, a temperature of 260°C, and a copper-based catalyst are used.

$$2H_2 + CO \longrightarrow CH_3OH$$

Methanol is used for various methyl compounds—dimethyl terephthalate for polyester fibers, methyl methacrylate, methyl acrylate, dimethyl phthalate, methyl chloride, and methylamines. It is also used in the United Kingdom in a small operation as a substrate for the growth of microorganisms that are harvested for use as a high protein animal feed. Imperial Chemical Industries is building a full-scale plant due for completion in 1980.

Methanol's largest outlet however is as a source of formaldehyde. Of the 6.5 billion lb of methanol made in the United States in 1977, 2.5 billion lb were consumed in this way. The oxidation is facile. In the high pressure process methanol and air are passed at 270 atm and 300°C over a zinc-chromium oxide catalyst. The low pressure process, which is becoming the more popular even though it is not yet predominant in the United States, operates at 50 atm and 250°C over a copper catalyst. Several medium pressure processes are also available, and a medium pressure plant is operating in Japan.

$$CH_3OH + \tfrac{1}{2}O_2 \longrightarrow HCHO + H_2O$$

$$CH_3OH \longrightarrow HCHO + H_2$$

About 50% of formaldehyde production goes into the large market for phenol-formaldehyde and urea-formaldehyde resins and the smaller market for the premium melamine-formaldehyde materials. Formaldehyde is also the raw material for polyacetal resins of the "Delrin" and "Zelcon" type (Sec. 4.2). Polyacetals are engineering plastics and are made in the United States, Germany, and Japan.

Formaldehyde is the basis for pentaerythritol (Sec. 2.4.3) and can combine with acetylene to give butynediol and a series of related chemicals (Sec. 2.11). With ammonia it yields hexamethylene tetramine.

$$6HCHO + 4NH_3 \longrightarrow \quad + \quad 6H_2O$$

Hexamethylene tetramine

"Hexa" is an intermediate in the manufacture of RDX, the explosive that replaced TNT in "blockbuster" bombs in World War II and the Korean War. Its principal peacetime use is as a convenient source of formaldehyde under alkaline conditions for the curing of phenolic resins (Part II, Sec. 2.3.1).

2.12.3 REACTIONS OF CARBON MONOXIDE

The third important use of synthesis gas is in the oxo process (Secs. 2.5.6; 5.5). Almost any primary alcohol may be made by the oxo process from the appropriate olefin. Those that are made on the largest scale are n-butanol together with some isobutanol from propylene (Sec. 2.5.6), isooctanol from heptene, decanol from nonene, and tridecanol from dodecene. In the United Kingdom a mixture of C_6—C_8 olefins obtained from slack wax is converted to a C_7—C_9 primary alcohol mixture known as "Alfanol" which replaces 2-ethylhexanol in many applications. Most oxo alcohols are used as their phthalates to plasticize PVC (Part II, Sec. 11.2).

Carbon monoxide itself undergoes a number of important reactions. It will react with methanol over an iodine-supported rhodium catalyst to give acetic acid (Sec. 2.4.3). In a similar process it reacts with dimethylamine to give dimethylformamide.

$$CH_3 \diagdown \!\!\!\!\!\diagdown NH + CO \longrightarrow CH_3 \diagdown \!\!\!\!\!\diagdown N\text{—}\overset{\displaystyle O}{\overset{\|}{C}}\text{—}H$$

Dimethylformamide

Under development but not yet commercial is a process for combining H_2 and CO to yield ethylene glycol (Sec. 2.4.5). With chlorine, carbon monoxide gives phosgene.

$$CO + Cl_2 \longrightarrow COCl_2$$

The reaction is carried out over activated charcoal at 250°C. About 85% of US phosgene is consumed in the production of tolylene diisocyanate (Sec. 2.8).

2.13 SUBSTITUTE NATURAL GAS (SNG)

In all likelihood world reserves of natural gas will be depleted sooner even than those of oil and long before those of coal. That prospect, together with the current US shortage of natural gas, has led to an interest in synthetic methane known as substitute natural gas (SNG).

SNG can be obtained from petroleum fractions by a variant of steam reforming.

$$4C_n H_{2n+2} + (2n-2)H_2O \longrightarrow (3n+1)CH_4 + (n-1)CO_2$$

The Catalytic Rich Gas (CRG) process (450°C, potassium promoted nickel catalyst) and the Gas Recycle Hydrogenator (750°C, no catalyst) have both been in operation for many years. The former, together with a variant called the "double methanation process," requires a light naphtha feedstock whereas the gas recycle hydrogenator can use heavier hydrocarbons. A process under development, called the "fluidized bed hydrogenator," will be able to operate on crude oil. As almost half the oil barrel is residual oil, this would be a major advance.

In the long term, however, the aim must be to obtain SNG from coal. One route, which can also be made to give a petroleum like liquid, is shown in Section 3.1.2. It goes by way of synthesis gas, which can undergo the reverse of steam reforming. All chemical reactions are reversible in principle, but steam reforming also works in practice. There is little difficulty in reforming mixtures of carbon monoxide, carbon dioxide, and hydrogen to methane, and the problem area is the gasification of coal.

The difficulty is one of solids handling. The supply of steam and oxygen in a molar ratio between $2:1$ and $6:1$ to a bed of coal leads to a mixture of hydrogen, carbon monoxide, carbon dioxide, and a little methane plus a mass of clinker (fused ash). The gases can be reformed, but the clinker is difficult to handle in a continuous plant. In the Lurgi process, developed in Germany in the 1930s, a high ratio of steam to oxygen was used which kept the temperature down and produced a gas low in carbon monoxide and a fine ash that could be handled relatively easily. The excess steam is expensive and leads to large amounts of dilute but corrosive effluent. It also leads to increased plant size for a given capacity and a low thermal efficiency.

A recent British process—the Westfield Slagging Gasifier—involves a low steam:oxygen ratio that gives a high proportion of carbon monoxide in the gas plus a slag that, at the high temperatures achieved, can be drawn off as a liquid. The US Energy Research and Development Administration commissioned the building of a demonstration plant in 1977.

2.14 CONCLUSIONS

We have now completed our discussion of the chemistry of the raw materials that come directly from petroleum and natural gas. We have stressed continually the high level of ingenuity that has made possible the production of the huge range of modern industrial organic chemicals from only seven basic chemicals. Ethylene, propylene, and benzene are the "biggies," and methane is also important although much of it is used for the inorganic chemical, ammonia. Butylenes, toluene, and xylenes are used on a substantial but smaller scale, and there are a number of much lower volume materials such as the pentenes and higher olefins that we have ignored for reasons of space. There are also many details of the "biggies" that would be inappropriate in a book of this size, and we refer readers to the bibliography.

What conclusions can we draw from our overview of the petrochemical industry? The first is that it is not going to be easy to change the raw materials base for organic chemicals if natural gas and petroleum supplies become scarce. This is partly because the basic molecules are so simple and partly because a sophisticated and economically workable technology allows these simple raw materials to be used effectively.

There are, of course, many other routes to organic chemicals if petroleum and natural gas were truly not available. But none of the raw materials can be pumped out of the ground. They require mining, which is labor intensive and involves inconvenient and expensive solids handling, or they require planting, cultivation, and harvesting. Thus any alternative we can now visualize will not be as simple or as economical as the feedstocks we have at present, and their conversion to usable materials will require huge capital investments.

A second point is that the chemical industry uses a relatively small amount of petroleum and natural gas. In the United States the figure is between 6 and 6.5% of the total. Because of the high dependence on methane and ethane about 10% of natural gas is used, a smaller proportion coming from petroleum to give the 6 to 6.5% average. In the United Kingdom and West Germany the proportion is higher—over 10%—and in West Germany it is expected to rise to 15% by 1980. Even that is less than one-sixth of the oil barrel. Thus, although the chemical industry competes for hydrocarbons with those who use them as fuel it consumes such a small proportion of the total that any reasonable planning should leave chemicals very much in the picture. This also follows if one considers the size of the chemical industry, the number of jobs it provides, the high value added in its processes (Sec. 2.2), and the all-pervasiveness of its products. The organic chemical industry is a key supplier to many other large industries such as plastics, textiles, rubber, paints, pharmaceuticals, cosmetics, food, transportation, and paper.

A third point—perhaps the most important to the student for whom this book is written primarily—is that industrial organic chemistry is a sophisticated and subtle body of knowledge that has been built up by chemists as they have learned to change simple raw material molecules into complex molecules that can be sold in the marketplace. This chemistry is quite different from the "classical" organic chemistry described in textbooks and taught in schools. The "classical" organic chemical synthesis is usually a multistage process perhaps involving an ingenious and mechanistically explicable rearrangement and an exotic and highly selective reagent. The industrial organic chemist, on the other hand, is looking for the simplest way a reaction can be made to occur. He aims to pass a readily available feedstock (perhaps not even a pure material) through a hot tube together with a simple, cheap reagent. His ingenuity is expended on finding a selective catalyst that will carry out several chemical changes simultaneouly under appropriate conditions of temperature and pressure. Yields are important in continuous processes but not desperately so, for one can recycle unchanged reactants and by-products. Knowing the mechanism is a bonus and often a help; indeed no one can deprecate the importance of mechanistic understanding. But the industrial chemist, much as he may yearn for it, has proved that he can survive without a solid theoretical basis. Catalysis is the mainstay of the organic chemicals industry and is one of the least understood branches of chemistry (Chap. 5). Humorists have referred to it as a branch of applied witchcraft, but recent research shows some signs of unraveling its secrets.

An understanding of catalysis could provide a feedback from petroleum chemistry into more traditional spheres of organic chemistry. More than anything else, the chemist who wishes to capitalize on this needs to have understanding for and appreciation of catalysis—a genuine "feel" for the subject. This is not easy to acquire. Much of what information exists is tucked away in patents. But a knowledge of modern catalyst technology could be a vital adjunct to the basic capabilities that a chemist brings to his job.

REFERENCES AND NOTES

Chemicals from natural gas and petroleum occupy much space in the general books on the chemical industry referred to in Sec. 0.4.2. Among the books devoted specifically to the petrochemical industry, a classic is A. L. Waddams, *Chemicals from Petroleum*, 4th ed., John Murray, London, 1978 (available from John Wiley, New York). A. V. Hahn, *The Petrochemical Industry—Production and Markets*, McGraw-Hill, New York, 1970, has a unique collection of data on manufacturing costs and markets in 1966, now unfortunately dated. R. B. Stobaugh, *Petrochemical Manufacturing and Marketing Guide*, 2 vols., Gulf Publishing Co., Houston, Texas, 1966, 1968, is strong on markets but is also dated. Stobaugh is said to have another book nearing completion. A useful paperback is J. P. and E. S. Stern, *Petrochemicals Today*, Arnold, London, 1971. *The Encyclopedia of Chemical Technology* has a great deal of information on petrochemicals and petroleum refining. Of importance to the serious student is *Hydrocarbon Processing,* a journal whose articles provide great understanding of the petroleum and petrochemical industries.

Sec. 2   The chemical industry's response to more expensive and less available natural gas is described in *Chem. Week,* August 11, 1976.

Sec. 2.1   The world oil industry, although not a theme of this book is of importance to anyone concerned with petrochemicals. The best introduction is *Our Industry— Petroleum,* British Petroleum, London, 1970. Two nontechnical books for general reading are A. Sampson, *The Seven Sisters,* Hodder and Stoughton, London, 1975, which recounts the history of the seven great oil companies, and C. Tugendhat and A. Hamilton, *Oil—The Biggest Industry,* 2nd ed., Eyre Methuen, London, 1975.

Statistics on world oil production and consumption are published annually in the *BP Statistical Review of the World Oil Industry,* BP, London, a well presented publication with the added advantage of being free. The Shell International Petroleum Co. also publishes a useful *Information Handbook,* London, 1977–1978, and a *Chemicals Information Handbook,* London, 1977.

The situation regarding lead in gasoline is not consistent throughout the world. At present new cars in the United States must run on lead-free gasoline. In continental Western Europe, lead levels have been reduced to about 0.15 g liter^{-1} while in the United Kingdom, where there is much skepticism about the evils of lead and the necessity for afterburners, levels still average around 0.5 g liter^{-1}. The British view is supported in an article in *Chem. Eng. News,* February 10, 1975, p. 13. Typical of articles describing the effect of lead removal from gasoline on the petrochemical industry is one in *Chem. Week,* Sep. 15, 1976, p. 11 and *Bus. Week,* Jan. 31, 1977, 38. A substitute for lead compounds now said to be included in some gasoline in the United States is methylcyclopentadienyl manganese tricarbonyl. Ecologists are questioning the effect of manganese compounds in automobile exhaust on the environment.

An old book on the chemistry of petroleum, which has never really been superseded, is B. T. Brooks, C. T. Boord, S. S. Kurtz, and L. Schmerling, *The Chemistry of the Petroleum Hydrocarbons,* 3 vols., Reinhold, New York, 1955.

The 1970 figures in Table 2.2 are based on P. Collinswood, *Ethylene in Europe,* Proc. of AIChE 68th. Ann. Mtg., March 1971, Houston, TX. Waddams *op. cit.* p. 26 gives different figures which nonetheless show the same trend. The 1976 figures come from the *Oil and Gas Journal,* 22 Nov. 1976 p. 79.

Sec 2.2.1   Historical insight into the development of the cracking process is described in an article in *CHEMTECH,* March, 1976, p. 180. The article reproduces William Burton's acceptance address for the 1922 Perkin Medal which was awarded him for his research on cracking.

Another article describing early work on cracking and the development of the petrochemical industry generally has been published by B. Achillaides, *Chem. Ind.,* April, 19, 1975, p. 337. The Burton process for cracking was first, but close on its heels was the Dubbs process. Dubbs was so involved with technology that he named his son Carbon Petroleum and his daughters, Methyl and Ethyl.

For descriptions of gas oil cracking, see a series of articles by S. B. Zdonik in *Hydrocarbon Proc.*, September 1975; December 1975; April 1976. Gas oil cracking is also discussed by M. J. Offen, *Hydrocarbon Proc.*, October, 1976, p. 123.

An excellent description of the development of catalytic reforming has been published by M. J. Sterba and V. Haensel, *Ind. Eng. Chem., Prod. Res. Dev.*, **15**, No. 1, (1976) 2.

For articles describing the cracking of whole oil see *Eur. Chem. News*, **30**, No. 785 (1977); *Chem. Week*, September 28, 1977, p. 39; December 7, 1977, p. 55. A discussion of substitute feedstocks by the year 2000 will be found in *Chem. Eng. News*, September 26, 1977, p. 7.

Table 2.3 is derived from Hatch and Mattar (see notes to Section 2.3 below) parts 8 & 9 January and March 1978.

Sec. 2.2.2 For the mechanism of thermal cracking see P. A. Wiseman, *J. Chem. Ed.*, **54**, No. 154 (1977).

Sec. 2.2.4 A review of octane number has been published by J. Benson, *CHEMTECH*, January 1976, p. 16.

Sec. 2.2.5 Use of hydrocracking on residues from petroleum distillation as well as coal-derived liquids, shale oil, and tar sands is described in an article in *Chem. Week*, February 18, 1976, p. 69.

Sec. 2.3 The changing feedstock patterns in the United States and United Kingdom is of absorbing interest to commentators. See for example P. Collinswood, *Ethylene in Europe*, Proc. of AIChE 68th Ann. Mtg., March 1971, Houston, TX: J. R. Lambrix and C. S. Morris, *Petrochemical Feedstocks for North America*, *Chem. Eng. Prog.*, August 1972, p. 24; J. R. Lambrix and C. S. Morris, *Petrochemical Feedstocks for North America—Gas and/or Liquids*, AIChE Symp., Series **69**, No. 127 (1973); *Chem. Eng. News* August 9, 1976, p. 13, November 15, 1976, p. 11; *Eur. Chem. News*, April 16, 1976, p. 4; K. Stork, M. A. Abrahams, and A. Rhoe, *More Petrochemicals from Crude*, *Hydrocarbon Proc.*, November 1974, p. 158; *Eur. Chem. News* February 10, 1978; *Chem. Age*, April 14, 1978; and a series of articles, "Hydrocarbons to Petrochemicals," by L. F. Hatch and S. Mattar in *Hydrocarbon Proc.* which started in May 1977. Union Carbide, Dow, and the Japanese are devising processes in which crude oil is cracked in a hot gas stream. There is little European interest in crude oil cracking. The feedstock situation is also discussed in *Propane Feedstocks in Petrochemical Plants*, Proc. of AIChE 79th Nat. Mtg., March 1975, Houston, Texas.

Sec. 2.4 The supply situation for ethylene through 1980 is described in an article in *Chem. Eng. News*, April 4, 1977, p. 9.

A detailed article about the consumption of ethylene and the other basic petrochemical feedstocks is contained in *Chem. Mktg. Rep.*, March 28, 1977, p. 29. Other articles dealing with ethylene production, the shift to heavier feedstocks, and the effect of this shift on coproducts are found in *Chem. Eng.*, March 28, 1977, p. 63; January 5, 1976, p. 116.

Plans to produce olefins at the major sources of petroleum such as the Middle East and Nigeria are well along. One proposal is that these olefins then be brought by refrigerated tanker to the countries where they are needed for conversion to downstream petrochemicals. This possibility is described in an article in *Chem. Week*, December 8, 1976, p. 44.

On the boards for 1980 is an olefins plant to be built by Shell Chemical that may be typical of plants of the future. Its design capacity will make possible the production of 1.5 billion lb of ethylene, 1 billion lb of propylene, and 500 million lb of butadiene. The plant at current cost levels is estimated to require $500 million worth of capital investment. It is claimed that it will not create an overcapacity of ethylene because current plants with capacities of 500–800 million lb per year will be uneconomical by the 1980s and will have to close. The Shell plant will operate on heavy gas oil. The boiler fuel will be fuel oil and pitch produced by the cracking process. However, conversion to coal as boiler fuel is contemplated.

The oligomerization of ethylene is described in US patents 3,644,563, 3,676,523, and 3,737,475.

The standard work on ethylene is S. A. Miller, Ed., *Ethylene and its Industrial Derivatives*, Benn, London, 1969. A particularly good review of the changes since 1973 is given in T. B. Baba and J. R. Kennedy, "Ethylene and its Coproducts: The New Economics," *Chem. Eng.*, January 5, 1976, p. 116.

Sec. 2.4.1   Oxychlorination technology has been described in detail in an article in *Chem. Week*, August 22, 1964, p. 93 and by Allen Clark, *Rev. Pure Appl. Chem.*, **21**, (1971) 145.

Sec. 2.4.3   For further discussion of the mechanism of the Wacker process for ethylene oxidation to acetaldehyde, see D. R. Armstrong, R. Fortune, and P. G. Perkins, *J. Catal.*, **45**, (1976) 339 and *CHEMTECH*, July, 1977, p. 398. The Wacker process for acetaldehyde production from ethylene is described by Jira, Blau, and Grimm, *Hydrocarbon Proc.*, March, 1976, p. 97.

The mechanism of the rhodium complex catalyzed carbonylation of methanol to acetic acid is described in an article by D. Forster, *J. Am. Chem. Soc.*, **98**, (1976) 846.

That choice of a process depends on local conditions known only to the manufacturer is emphasized by a note in *Chem. Week*, April 7 1976, p. 39 that indicates that Celanese is increasing capacity to make acetic acid from acetaldehyde, which in turn is obtained from the hydration of ethylene. They are also maintaining their plant for producing acetic acid by the liquid phase oxidation of butane, and they are building still another plant to make acetic acid by the carbonylation of methanol.

Sec. 2.4.4   US styrene capacity and demand has been discussed by T. C. Ponder, *Hydrocarbon Proc.*, July 1977, p. 137. New styrene plants with decreased energy usage have been described in an article in *Chem. Week*, November 17, 1976, p. 34.

Insight into the production of ethylbenzene is provided in an article in *Chem. Week*, June 4, 1975, p. 29.

Sec. 2.4.5   The preparation of ethylene glycol by chlorohydrin technology, ethylene oxidation, and acetoxylation has been compared by A. M. Brownstein, *Chem. Eng. Prog.*, **71**, No. 9, (1975) 72. He also includes the route from CO and hydrogen.

Dutch patents 07383 (1974) and 07412 (1974) describe processes for conversion of synthesis gas to ethylene glycol.

Sec. 2.4.6   A description of Union Carbide's 150 million lb $year^{-1}$ propionaldehyde plant and a discussion of uses for propionaldehyde is contained in an article in *Chem. Eng. News*, October 13, 1975, p. 6. The carbonylation of ethylene to propionic acid is described in US patent 3,944,604.

Sec. 2.5   E. G. Hancock, Ed., *Propylene and Its Industrial Derivatives*, Benn, London, 1973, is a good reference for olefins. The market situation is given by P. H. Spitz, "Propylene: Future Supply/Demand," *Hydrocarbon Proc.*, July 1976, p. 143 and in "Propylene Through the Seventies," *Chem. Eng. Prog.*, December, 1972, p. 61.

Sec. 2.5.3   The preparation of methyl methacrylate by the oxidation of isobutene is discussed in an article by Oda et al., *Hydrocarbon Proc.*, October 1975, p. 115.

Sec. 2.5.4   The simultaneous preparation of propylene oxide and methyl methacrylate is described in Japanese patent 43,926 (1974).

Sec. 2.5.5   Synthetic routes to glycerol from propylene are described in an article by K. Yamagishi, *Chem. Econ. Eng. Rev.*, **6**, (1974) 40.

Sec. 2.5.6   The modification of the classical cobalt carbonyl catalyst approach to hydroformylation to increase the yield of linear aldehydes and to allow the use of lower temperatures is described by Kummer and co-workers in D. Forster and J. F. Roth, *Homogeneous Catalysis—II*, American Chemical Society, Washington, DC, 1974, Chapter 2, pp. 19–27. This same volume contains several other chapters describing recent hydroformylation technology.

    The commercialization of the rhodium hydroformylation of olefins is described in an article *Chem. Eng. News*, Apr. 26, 25 (1976). Carbonylation with rhodium is also described in *CHEMTECH*, December 1976, p. 772.

    The mechanism of hydroformylation utilizing rhodium catalysts is contained in an article *Chem. Eng.*, December 5, 1977, p. 112.

Sec. 2.5.7   Numerous articles have appeared on metathesis. A good review has been published in *CHEMTECH*, August 1975, p. 486. The mechanism of the reaction has been discussed by Muetterties, *Inorg. Chem.*, **14**, (1975) 953 and by Katz and McGinnis, *J. Am. Chem. Soc.*, **97**, (1973) 1592. Additional articles of interest are by K. L. Anderson and T. D. Brown, *Hydrocarbon Proc.*, August 1976, p. 119; R. L. Banks, *Fortschr. Chem. Forsch.*, **25**, (1972) 39; G. C. Bailey, *Catal. Rev.*, **3**, (1969) 37; and W. B. Hughes in D. Forster and J. F. Roth, *op. cit.*

Sec. 2.6.1   For more information on butadiene production and markets see T. Reis, *Butadiene Extraction*, *Chem. Proc. Eng.*, March 1970, p. 65; W. K. Hayes, *Specialty Uses of Butadiene*, *Chem. Eng. Prog.*, Symposium Series, **66**, (1970) 103, 60; *Hydrocarbon Proc.*, August 1972, 102; R. L. Ericsson, "Butadiene Outlook," *Hydrocarbon Proc.*, November 1974, p. 158; and *Eur. Chem. News*, October 1975, pp. 17, 19.

Sec. 2.6.2   Patents relating to the production of adiponitrile from butadiene by the direct addition of HCN include Ger. Offen. 2,221,137 (1972); Ger. Offen. 2,225,732 (1972); US patent 3,655,723 (1972); French patent 2,069,411 (1971); Ger. Offen. 1,817,800 (1971); Ger. Offen. 1,817,797 (1971); US patent 3,542,847 (1970); and US patent 3,551,474 (1970).

    The hydrocyanation of monoolefins has been reviewed by E. S. Brown in "*Aspects of Homogeneous Catalysis*," R. Ugo, Ed., Vol. 2, D. Riedel, Boston, p. 57.

    In spite of the favorable economics of the HCN/butadiene route to adiponitrile, the hydrodimerization route still flourishes and a case history, *Discovery, Development and Commercialization of the Electrochemical Adiponitrile Route* by M. M. Baizer and D. E. Danly, appeared in *Chemistry & Industry*, July 7, 1979, p. 435.

Sec. 2.6.3   A comparison of the processes for manufacturing maleic anhydride from butane and from benzene is contained in an article in *Chem. Week*, October 13, 1976, p. 79.

Articles relating to C_4 olefin supply in the future are contained in *Chem. Week*, September 8, 1976, p. 70; *Chem. Eng. News*, September 13, 1976, p. 11; and *Chem. Week*, March 2, 1977, p. 31; November 16, 1977, p. 49.

Sec. 2.7   A review of benzene supply and demand has been published in *Chem. Eng. News*, May 10, 1976, p. 13. The US benzene supply and demand situation has also been reviewed by J. E. Fick, *Hydrocarbon Proc.*, **55,** (1976) 127.

E. G. Hancock, Ed. *Benzene and Its Industrial Derivatives*, Benn, London 1975 is a standard work.

Sec. 2.7.1   The standard cost comparison for phenol processes still appears to be that of A. S. Biancu, *Chem. Proc. Eng.*, January 1967. The matter is attractively discussed in *Chemistry Case Studies* cited in Sec. 0.4.3.

Sec. 2.7.3   The amination of benzene is described in US patents 4,031,106 (1977) and 3,919,155 (1975).

Aniline supply and demand has been described in *Chem. Eng. News*, September 9, 1974, p. 7.

Sec. 2.8   US patent 3,980,580 describes the oxidative coupling of toluene to stilbene and stilbene's subsequent metathesis with ethylene to styrene.

The process for obtaining styrene from toluene by way of stilbene has also been described in *CHEMTECH*, March 1977, p. 140.

Sec. 2.9   The market situation for xylenes discussed in W. A. McCormick and V. A. Bonanni, *Xylenes-Supply/Demand Next Ten Years*, AIChE Symp. Series **69,** (1973), 127, 176. The discovery of a new catalyst that led to increased plant capacity, combined with a decline in consumer interest in double knit fabrics based on polyester fibers, made for a surplus of *p*-xylene in 1977. For the first time in history its price dropped below that of *o*-xylene.

Sec. 2.11   A review of acetylene written in its heyday is S. A. Miller, *Acetylene, Its Properties, Manufacture and Uses*, Benn, London, 1965.

Sec. 2.12   Production of ammonia and hydrogen and the steam reforming process are discussed in *Catalyst Handbook*, Imperial Chemical Industries, Wolfe, London, 1970. Production of synthetic methane (substitute natural gas or SNG) is reviewed by D. Rooke, "Innovation in the Gas Industry, " *Chem. Eng.*, January 1978, p. 34. Gas processing is a special topic in each April issue of *Hydrocarbon Proc.*, and the various SNG processes are compared in the April 1975 issue.

Sec. 2.1.2   A new process for converting methanol to gasoline with a zeolite catalyst has received considerable publicity. Articles describing this process can be found in *Chem. Eng. News*, January 30, 1978, p. 26 and *CHEMTECH*, February 1976, 86.

The conversion of methanol to ethanol is a distinct possibility using complex cobalt catalysts with phosphine ligands. This is described in Dutch patent 06138 (1976).

Chapter Three
SOURCES OF CHEMICALS OTHER THAN NATURAL GAS AND PETROLEUM

In Chapter 2 we described the derivation of chemicals from petroleum. But organic chemicals can come from other sources—coal, fats, oils, and carbohydrates. Historically these sources are important because it was from them that the modern chemical industry evolved. The majority of the world's ethanol is still made by fermentation even though industrial ethanol (as opposed to beverage ethanol) has been made by hydration of ethylene since about 1950. In a pinch we could make ethylene by dehydration of fermentation ethanol as we did before World War II.

The traditional sources of fermentation ethanol are carbohydrates such as grain and potatoes. Whether we want to divert these foodstuffs to chemical uses is a point open to practical and philosophic argument. (A converse point is that petrochemicals have set free quantities of foodstuffs for which the chemical industry would otherwise have competed.) Certainly it would be costly to make ethylene from fermentation ethanol, but it has one great advantage—carbohydrate feedstocks are renewed annually and draw their energy from the sun, which is powered by nuclear fusion and as an energy source appears viable for a good many years to come.

We are interested in fats, oils, and carbohydrates, therefore, not only because of their present applications, but also because they represent an insurance policy for the future. Coal, although a nonrenewable source of chemicals and energy, occurs on earth in much larger quantities than petroleum and will certainly outlast petroleum reserves by a few hundred years.

3.1   COAL

Chemicals can be obtained from coal mainly in four ways:

1 From the coal tar distillate that results from the conversion of coal to coke.
2 By conversion of coke to water (synthesis) gas from which methane and Fischer-Tropsch hydrocarbons can be obtained (see Notes).
3 By hydrogenation of coal to yield methane or petroleumlike substances.
4 By conversion of coal to acetylene (Sec. 2.11) by way of calcium carbide.

3.1.1   COAL TAR CHEMICALS

When coal is heated in the absence of air to a temperature of about 1000°C coke forms together with a number of liquid and gaseous decomposition products. The coke is almost pure carbon and is used in steel manufacture. As steel manufacture becomes more efficient, however, the need for coke decreases. Nonetheless, it will always have a market, and therefore we shall always have available the chemicals that volatilize from the coke ovens. These chemicals include benzene, toluene, xylenes, naphthalene, coumarone, indene, and phenol. Anthracene, cresols, xylenols, pyridine, and many other minor constituents are produced in smaller amounts. Benzene is the major constituent of coal tar but still accounted for only 4.4% of the huge US benzene consumption in 1977. Most important in market terms is naphthalene, which is a raw material for phthalic anhydride in competition with *o*-xylene (Sec. 2.9). Until the early 1960s coal tar was the sole source of naphthalene, but it can now be obtained by catalytic reforming of heavier naphthas. Unfortunately, methylnaphthalenes, form, and the methyl groups must be removed by hydrodealkylation if naphthalene itself is to be obtained. This is the same reaction used to convert toluene to benzene (Sec. 2.8).

The reliance of the chemical industry on coal-based naphthalene is shown by the fact that of the 800 million lb of naphthalene produced in the United States in 1976, 55% came from coal tar. No other coal-based chemical is of such importance. The quantities of coal tar available are tied to the production of steel and could not be increased substantially if other sources of chemicals diminished.

3.1.2 THE FISCHER-TROPSCH REACTION

The production of hydrogen/carbon monoxide mixtures from methane or naphtha and their uses are described in Section 2.10. They also result when coke reacts with steam:

$$C + H_2O \rightleftharpoons CO + H_2$$

The reaction is highly endothermic, and accordingly the bed of coke is first heated by the burning of a portion of it in air:

$$C + O_2 \rightleftharpoons CO_2$$

or

$$C + \tfrac{1}{2}O_2 \rightleftharpoons CO$$

This process is highly exothermic, and when the coke has reached white heat the air flow is replaced by water. When the coke has cooled again the cycle is repeated. The mixture is also enriched with hydrogen by the water gas reaction:

$$CO + H_2O \rightleftharpoons CO_2 + H_2$$

Hydrogen enrichment is important for production of both methanol and methane. The reaction of carbon monoxide and hydrogen to give methane (the reverse of steam reforming) has already been mentioned (Sec. 2.12). This methanation reaction is one possible route from coal to substitute natural gas (Sec. 2.13). The important reactions are

$$\mathbf{a} \quad CO + 3H_2 \rightleftharpoons CH_4 + H_2O$$

$$\mathbf{b} \quad CO_2 + 4H_2 \rightleftharpoons CH_4 + 2H_2O$$

$$\mathbf{c} \quad 2CO \rightleftharpoons C + CO_2$$

$$\mathbf{d} \quad CH_4 \rightleftharpoons 2H_2 + C$$

Reactions **c** and **d** are troublesome because the carbon produced can foul the catalyst. Excess steam reduces carbon formation although it also depresses slightly the yields of methane.

Methanation catalysts include several transition elements of which nickel on silica is particularly effective. A plant in South Africa has demonstrated the feasibility of methanation and a great deal of related work is underway worldwide.

The Fischer-Tropsch reaction is another interesting route from coal to hydrocarbons. When synthesis gas is passed over an iron, nickel, or cobalt catalyst at 150–300°C and at near-atmospheric pressures, a mixture of alkanes and alkenes with a broad range of molecular weights is formed. The hydrocarbons are predominantly C_5–C_{11} straight chain, and there are also some oxygenated compounds such as alcohols and acids. The result is a petroleumlike mixture that can be used both as a fuel and a petroleum feedstock. A small amount of wax is obtained.

Ruthenium has been proposed as a catalyst to yield higher molecular weight hydrocarbons. The reactions involved include the following:

$$nCO + (2n+1)H_2 \longrightarrow C_nH_{2n+2} + nH_2O$$

$$nCO + 2nH_2 \longrightarrow C_nH_{2n} + nH_2O$$

$$2nCO + nH_2 \longrightarrow C_nH_{2n} + nCO_2$$

Since the hydrocarbons are straight chain they have low octane numbers and must be isomerized for use, for example, in the alkylation reaction (Sec. 2.2.4). On the other hand, the straight chain structure is ideal for steam cracking and catalytic reforming to provide the olefins and aromatics needed for chemical synthesis.

The process was subject to intensive development between the world wars and was operated successfully on a large scale in Germany during World War II. The only plant currently operating is in South Africa at Sasolburg. A large expansion is underway. The government, with an abundance of cheap coal, operates it as an insurance against its racial policies provoking sanctions and a petroleum boycott.

In spite of its superficial virtues the process is costly and unreliable in operation. Coke is a solid; consequently reactors are bulky, mechanically complex, and therefore expensive. Utility and maintenance costs are high, and coal has to be dug laboriously out of the ground.

We have described the Fischer-Tropsch reaction here rather than in Section 2.12 on synthesis gas because it does not seem sensible to make synthesis gas from petroleum and then reverse the process to make petroleum from synthesis gas. It is possible to imagine circumstances, however, under which that might be desirable. Petroleum-based synthesis gas is derived from methane or light naphtha. There are, however, many heavier petroleum fractions such as fuel oil and asphalt that are not used for gasoline or chemicals. If these could be converted cheaply to synthesis gas, which in turn could be reconverted to a lighter fraction suitable for gasoline or chemicals, the process might be worthwhile, at least in the United States.

3.1.3 COAL HYDROGENATION

Coal has the approximate empirical formula CH. To convert it to an aliphatic hydrocarbon mixture requires hydrogen. In the Fischer-Tropsch process this is derived ultimately from water. It is, however, possible to hydrogenate coal directly. This process, called the Bergius process, was operated in Germany in World War II. Coal, lignite, or coal tar was hydrogenated over an iron catalyst at 450°C and 700 atm pressure, and 4 million tons of gasoline were synthesized in this way.

There are thus two routes from coal to petroleumlike fuels. The basic technology of both is wellknown, and additional development work is in progress. Whether or not coal ever becomes a raw material for petroleumlike fuels on a world scale depends on the severity of the petroleum shortage, the availability of the huge capital investment required, and the feasibility of various other possible sources of energy such as nuclear fission or nuclear fusion.

3.1.4 CARBIDE

The production of acetylene by way of carbide and its uses have already been discussed (Sec. 2.11). The process uses energy extravagantly, and at 1978 US prices the cost of electricity per pound of acetylene is greater than the total cost of a pound of ethylene. It might become possible for carbide to again become interesting if nighttime electricity becomes cheap, because nuclear reactors prefer to operate steadily around the clock. Even then it seems intuitively that electrolysis of water would be a more sensible way to use the surplus electricity.

3.1.5 COAL AND THE ENVIRONMENT

Coal poses social and political as well as technical problems. The cheapest way to mine coal is by open pit or strip mining, which ruins large areas of countryside. On the other hand, traditional underground mining is a dangerous business and one that is felt by many to demean the human spirit. Mechanization of mining and better safety standards have improved the miner's lot from what it was in the nineteenth century, but old attitudes die hard. The production of coal at a lower social cost may be a problem of crucial importance for future generations.

3.2 FATS AND OILS

Naturally occurring fats and oils are triglycerides, that is esters of glycerol with saturated or unsaturated fatty acids. They may be of animal or

Formula	Name	Melting Point (°C)	Double Bond Position and Stereo Chemistry	Source
$n\text{-}C_{11}H_{23}COOH$	Lauric acid	44.2	—	Coconut oil, palm kernel oil
$n\text{-}C_{13}H_{27}COOH$	Myristic acid	53.9	—	Coconut oil, palm kernel oil
$n\text{-}C_{15}H_{31}COOH$	Palmitic acid	63.1	—	Most vegetable oils and animal fats
$n\text{-}C_{17}H_{35}COOH$	Stearic acid	69.6	—	Most vegetable oils and animal fats
$n\text{-}C_{17}H_{33}COOH$	Oleic acid[a]	16.0	cis-9	Most vegetable oils and animal fats (olives, nuts, beans), tall oil
$n\text{-}C_{17}H_{31}COOH$	Linoleic acid[a]	−9.5	cis-9, cis-12	Tall oil, most vegetable oils (saf-flower, sunflower, soy)
$n\text{-}C_{17}H_{29}COOH$	Linolenic acid[a]	−11.3	cis-9, cis-12, cis-15	Linseed oil
$n\text{-}C_{17}H_{29}COOH$	α-Eleostearic	48.5	cis-9, trans-11, trans-13	Tung oil
$n\text{-}C_{17}H_{29}COOH$	β-Eleostearic	71.5	trans-9, trans-11, trans-13	Tung oil
$n\text{-}C_{17}H_{35}(OH)COOH$	Ricinoleic acid[a]	5.0	cis-9	Castor oil

[a] Oleic acid: $CH_3(CH_2)_7CH{=}CH(CH_2)_7COOH$;
Linolenic acid: $CH_3(CH_2)_4CH{=}CH{-}CH_2{-}CH{=}CH(CH_2)_7COOH$;
Linolenic acid: $CH_3CH_2CH{=}CH{-}CH_2{-}CH{=}CH{-}CH_2{-}CH{=}CH(CH_2)_7COOH$;
Ricinoleic acid: (12-hydroxyoleic acid): $CH_3(CH_2)_4CH_2CHCH_2CH{=}CH(CH_2)_7COOH$.
$\qquad\qquad\qquad\qquad\qquad\qquad\qquad\quad |$
$\qquad\qquad\qquad\qquad\qquad\qquad\qquad\ \ OH$

Figure 3.1   Common fatty acids.

vegetable origin. They have the general formula: $ROOCCH_2$—$CH(OOCR)$—CH_2OOCR. Usually more than one species of acid is present, and thus the triglyceride is "mixed." Some of the fatty acids most commonly found in fats and oils are shown in Figure 3.1. They all have even numbers of carbon atoms.

Fats and oils are one of the three major groups of foodstuffs, the others being proteins and carbohydrates. Food is, however, surrounded with a host of cultural attitudes. In southern Europe oil is widely used for cooking whereas in northern Europe solid fats are preferred. In the Middle East a market for butter scarcely exists whereas in Europe and North America it is an important foodstuff. Therefore ways have been sought by which oils could be hardened to make them culturally accepta-ble to dwellers in the northerly areas. The melting point of fat or oil is

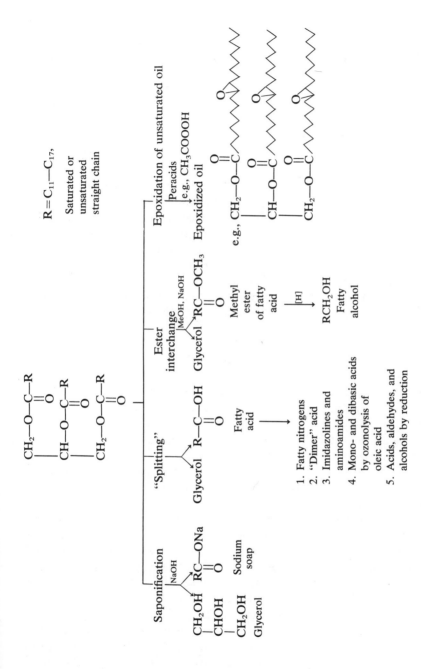

Figure 3.2 Chemical reactions of fats.

133

related to the melting points of the fatty acids it contains, and these in turn depend on molecular weight (see the series lauric, myristic, palmitic, stearic), number of double bonds (see the series stearic, oleic, linoleic, linolenic), and the proportion of groups arranged *trans* across the double bond (see the series linolenic, α-eleostearic, β-eleostearic). Crystalline structure of the fat also plays a role. To harden a fat or oil, therefore, it is hydrogenated and isomerized over a nickel catalyst. The degree of hardness can be controlled, and fats of any desired consistency can be produced (Part II, Sec. 13.1.1).

One of the most important commercial reactions of fats and oils that does not lead to foods or soaps is ester interchange or alcoholysis. If a triglyceride is heated with a polyhydroxy compound such as glycerol or pentaerythritol in the presence of a suitable catalyst, such as sodium methoxide, mixed partial esters are obtained. The partial esters may then be reacted with a dibasic acid or its precursor, usually phthalic anhydride, to give an oil-modified alkyd resin (Part II, Sec. 5.6). If the partial esters are reacted with tolylene diisocyanate, oil-modified urethanes are obtained. The other important chemical reactions of fats are shown in Figure 3.2.

3.2.1 FATTY ACIDS

Oils and fats are saponified to glycerol and soap by treatment with alkali (Fig. 3.2) when soap is the desired end product. Most widely used in both the United States and Europe to obtain free fatty acids is "splitting"— continuous, noncatalytic hydrolysis with high temperature and pressure. In smaller plants continuous autoclave splitting is used with oxide catalysts such as zinc oxide, and very small operations may use batch processes with so-called Twitchell catalysts, which are combinations of sulfuric and sulfonic acids.

The lower molecular weight acids, lauric and myristic, come from coconut and palm kernel oils (Fig. 3.1). Palmitic and stearic acids are found in most oils and fats, tallow being the most important. Oleic acid is a component of animal fats and is a major constituent of many vegetable oils. Linoleic acid is found in many vegetable oils, particularly linseed and safflower oil. 12-Hydroxyoleic acid or ricinoleic acid with its hydroxyl group is an oddity found in castor oil.

The most important fat or oil in the United States is soybean oil of which 12.1 billion lb were produced in 1975. This is about two-thirds of the world's production. Little of this, however, is used by the chemical industry which, although it could obtain fatty acids from triglycerides, obtains most of them from tall oil, which became important after World

War II. Tall oil is not a triglyceride but a mixture of rosin and fatty acids. Rosin comprises a complex mixture of monobasic acids related to partially hydrogenated phenanthrene. They are known as abietic, pimaric, and resin acids.

Tall oil is a by-product of the pulping of southern pine for paper manufacture. Southern pine is converted to pulp by the Kraft process in which sodium hydroxide is used to separate the desired cellulosic fibers from the undesired lignins (polymers of phenylpropane monomers containing OH and OCH_3 groups), rosins, fatty acids, and other materials in the wood. The fatty acids, mainly oleic and linoleic, and rosin end up as their sodium salts in a smelly black liquid. Acidification gives rosin and fatty acids, which may be separated by distillation.

The rosin is most famous for its use on violin bows, but far larger quantities go into elastomer formulations and paints and varnishes. As its sodium salt it is the important size in paper manufacture, and this is its largest use. More than half the world's supply of rosin is produced in the United States. European production is much smaller and comes mainly from France.

Another source of the vegetable oil fatty acids used by the chemical industry is "foots" or soapstocks. Vegetable oils tend to contain a small percentage of free fatty acids that result from enzymatic decomposition. If the oils are to be used as foodstuffs these must be removed. This is simply done with a small amount of alkali and the sodium salts of the fatty acids are then skimmed off. They are known as foots, and the fatty acids are regenerated by acidification. Associated with the fatty acids are small quantities of tocopherol, which is converted to Vitamin E, and small amounts of sterols, including stigmasterol, useful for conversion to cortisone (Part II, Sec. 8.8).

In Russia and Japan petrochemical-based fatty acids with mixtures of odd and even numbers of carbon atoms have been produced. This technology has not yet been adopted in Western Europe and is only now entering the United States. The products are considered inferior. However, short chain acids useful for synthetic lubricants (Part II, Sec. 10.10.1) are produced.

The large part of fatty acid production goes into soaps. Domestic soap is merely the sodium salt of the mixture of fatty acids obtained by hydrolysis of oils or fats (Fig. 3.2). The major components are sodium stearate and palmitate. The splitting of fats into fatty acids and glycerol was discovered by Chevreul in 1823, and the fatty acid mixture was called stearine. It is used not only for soap but also as a lubricant in rubber and polymer processing. In the nineteenth century, until the advent of cheap paraffin wax, it was the principle material used in household candles. It is still blended with paraffin wax in some candle formulations to improve the melting properties of the wax, but only in countries like Denmark and Sweden whose fishing industries produce a surplus of cheap stearine is it a major component of candles. Paraffin candles dominate the world candle market, which has now passed its nadir and is again growing. Paraffin wax comes from the heavier petroleum fractions, and ceresin or microcrystalline waxes from the same source are used in US candles. These are branched chain paraffins and consequently must be blended with an antioxidant to prevent oxidation at the tertiary carbon atoms. Fischer-Tropsch waxes are made in South Africa (Sec. 3.1.2); those of higher molecular weight are used throughout the world for the fashionable long, thin candles while the fraction equivalent to paraffin wax is used in the larger domestic market.

Salts of fatty acids are used as stabilizers for PVC (Part II, Sec. 2.2.1). These are the laurates or stearates of metals such as lead, barium, calcium, strontium, or zinc. Metal soaps of cobalt, lead, manganese, calcium, and zirconium can be added to unsaturated oils to accelerate their oxidation and are thus used as driers in oil-based paints. Metal soaps (e.g., of lithium) are also used to thicken lubricating oils to yield greases. Free fatty acids may be components of automobile lubricating oil formulations.

Oleic acid may be cleaved at the double bond by treatment with ozone from an electrical discharge. The products are azelaic and pelargonic acids, both of which are used in synthetic lubricants (Part II, Sec. 10.10.1).

$$CH_3(CH_2)_7CH{=}CH(CH_2)_7COOH \xrightarrow{O_3} CH_3(CH_2)_7COOH$$

Oleic acid   Pelargonic acid

$$+ HOOC(CH_2)_7COOH$$

Azelaic acid

3.2.2 FATTY NITROGEN COMPOUNDS

Fatty acids may be converted into a large number of fatty nitrogen compounds of which the fatty amines are the most important. Some of these are shown in Figure 3.3. They find many applications in industry as surface active agents, and over 100 million lb were produced in the United States in 1977.

In Figure 3.3 we see that the fatty nitrogen compounds start with fatty acids in which the alkyl group may contain 8–18 carbon atoms. Normally it is saturated although derivatives based on oleic acid are also produced. Treatment with ammonia converts the fatty acid to a nitrile through successive dehydration of the ammonium salt and the amide, neither of

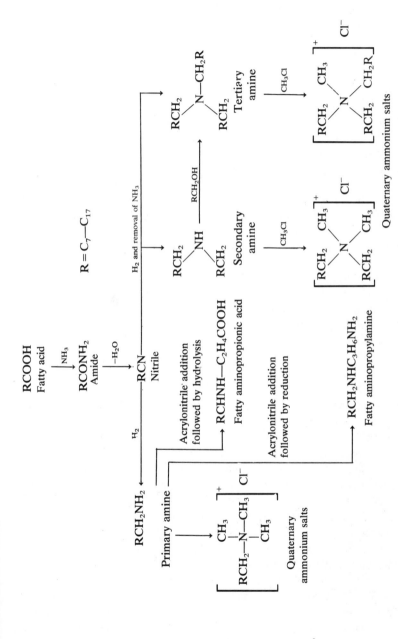

Figure 3.3 Fatty nitrogen chemistry.

which need be isolated. This chemistry was observed earlier in the classic synthesis of hexamethylene diamine (Sec. 2.6.2). Amides, however, are articles of commerce and are used as a parting agent or slip agent to prevent plastic from adhering to the die during extrusion. The nitrile in turn may be reduced to a primary, secondary, or tertiary amine. Conditions for primary amine formation require that ammonia be present to suppress secondary amine formation. Conversely, when the secondary amine is desired as a major product ammonia must be removed continuously from the reaction mixture. The tertiary amine may be prepared also by hydrogenation with removal of ammonia. It is better prepared, however, by the interaction of a di-(long chain alkyl) amine with an alcohol. The primary, secondary, and tertiary amines may be quaternized with methyl chloride or methyl sulfate. The primary amine may also be reacted with acrylonitrile to provide a fatty aminopropionitrile, which in turn can be reduced to a fatty aminopropylamine. Each of the compounds, as shown in Figure 3.3, varies in degree of surface activity, and it is this variation that accounts for their specific applications detailed in Part II, Section 7.2.

In addition to the foregoing simple reactions, fatty acids undergo a number of more complicated reactions of chemical interest, although the tonnages of materials involved are trivial compared with petroleum and natural gas. It is important to note, however, that the products are often not accessible from petrochemical sources.

3.2.3 "DIMER" ACID

The dimerization of linoleic acid (primarily from tall oil acids) is shown in **a** of Figure 3.4. Natural linoleic acid has double bonds in the 9,12 positions (Fig. 3.1), but when heated it isomerizes to the conjugated 10,12 or 9,11 structure. In **a** of Figure 3.4, the 9,11 acid is shown. This diene may then undergo a Diels-Alder reaction with another molecule of the original 9,12 acid or either of the conjugated isomers. Reaction **a** of Figure 3.4 shows the reaction with the 9,12 acid to give a typical Diels-Alder adduct—a cyclohexene with four side chains, two of which contain carbonyl groups.

The conjugation reaction can go in two ways, and the products can react with either of the double bonds in the 9,12 acid, the 10,12 acid, or the 9,11 acid. Furthermore, addition may take place head-to-head or head-to-tail. Reaction **a** of Figure 3.4 shows head-to-head addition. In all, 24 different products are possible. An added complication is that the double bonds might be *cis* or *trans*, although most naturally occurring double bonds including those in linoleic acid are *cis*. Whether they are *cis* or *trans*, however, affects markedly the kinetics of the reaction.

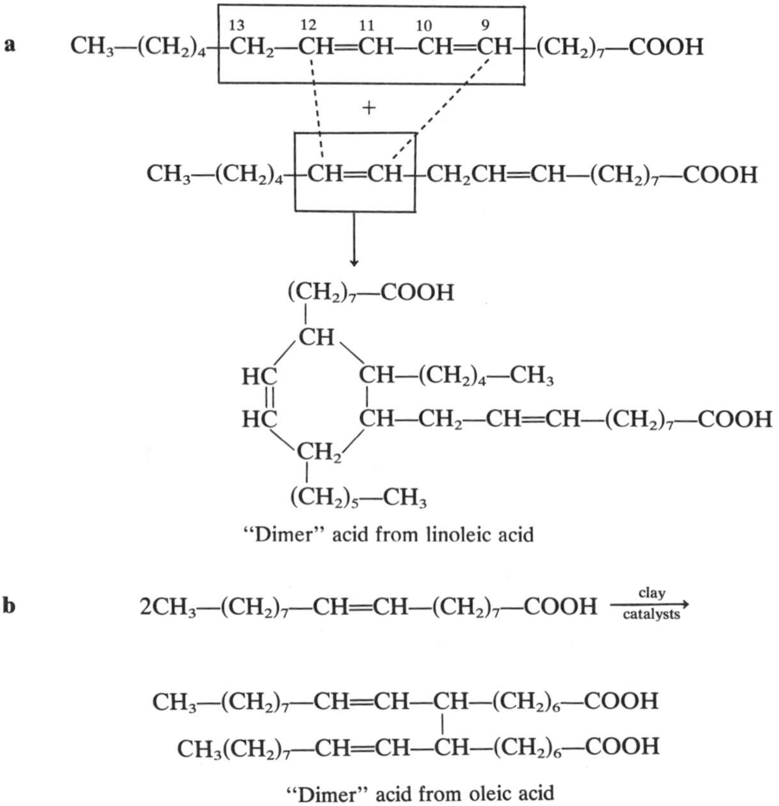

a $\quad CH_3$—$(CH_2)_4$—$\boxed{CH_2$—$\overset{13}{C}H$=$\overset{12}{C}H$—$\overset{11}{C}H$=$\overset{10}{C}H}$—$\overset{9}{(CH_2)_7}$—$COOH$

+

CH_3—$(CH_2)_4$—$\boxed{CH$=$CH}$—CH_2CH=CH—$(CH_2)_7$—$COOH$

$(CH_2)_7$—$COOH$

$\begin{array}{c} CH \\ HC \diagup \quad \diagdown CH\text{—}(CH_2)_4\text{—}CH_3 \\ HC \qquad \quad CH\text{—}CH_2\text{—}CH\text{=}CH\text{—}(CH_2)_7\text{—}COOH \\ \diagdown CH_2 \diagup \\ (CH_2)_5\text{—}CH_3 \end{array}$

"Dimer" acid from linoleic acid

b $\quad 2CH_3$—$(CH_2)_7$—CH=CH—$(CH_2)_7$—$COOH$ $\xrightarrow[\text{catalysts}]{\text{clay}}$

CH_3—$(CH_2)_7$—CH=CH—CH—$(CH_2)_6$—$COOH$
$CH_3(CH_2)_7$—CH=CH—CH—$(CH_2)_6$—$COOH$

"Dimer" acid from oleic acid

(The bond forms in many of the possible positions, and one double bond may be reduced)

Figure 3.4 "Dimer" acid.

Linoleic acid is relatively scarce and expensive, and it was necessary to find a way to dimerize the cheaper oleic acid that with its single double bond will not undergo the Diels-Alder reaction. If treated with a natural clay (a dehydrogenation catalyst dear to the heart of the petroleum industry) oleic acid will dimerize as shown in **b** of Figure 3.4. The product is known as "dimer acid," and chemically many different structures are present although all contain 36 carbon atoms and two carboxyl groups. These dibasic acids are used in the production of specialty polyamide resins with unusual adhesive and coating properties. For example, these are practically the only materials that adhere to polyethylene, and accordingly they are used as vehicles for printing inks for polyethylene film.

An interesting specialty use involves substituting dimer acid esters for lubricating oil in two cycle outboard engines in lakes in Switzerland that were being polluted by petroleum engine oil discharged from motor boats.

3.2.4   AMINOAMIDES AND IMIDAZOLINES

A small volume application of fatty acids involves their conversion to aminoamides and imidazolines. A fatty acid will react with a polyamine such as diethylenetriamine to give an aminoamide, a molecule of water being eliminated. Further dehydration leads to cyclization. Imidazolines have many of the properties of the fatty nitrogen compounds already described.

$$RCOOH \ + \ NH_2{-}C_2H_4{-}NH{-}C_2H_4{-}NH_2 \ \xrightarrow{-H_2O}$$

Fatty acid   Diethylenetriamine

$$\underset{\text{Aminoamide}}{R{-}\overset{\overset{\displaystyle O}{\|}}{C}{-}NH{-}C_2H_4{-}NH{-}C_2H_4{-}NH_2} \ \xrightarrow{-H_2O}$$

$$\underset{\text{Imidazoline}}{\begin{array}{c} R{-}C{=}N \\ | \qquad\quad \diagdown CH_2 \\ | \qquad\qquad | \\ H_2NC_2H_4{-}N{\rule{1em}{0.4pt}}CH_2 \end{array}}$$

If quaternized with methyl chloride or sulfate they are useful as textile softeners like the fatty quaternaries already discussed (Sec. 3.2.2). Their main use is as corrosion inhibitors particularly in oil well applications. They also find use as asphalt emulsifiers and asphalt antistrippants. An asphalt emulsifier is an emulsifying agent that brings particles of liquid asphalt and water into close proximity to provide a convenient means for laying down a layer of asphalt on a roadbed. An asphalt antistrippant causes asphalt to adhere to the rock or aggregate with which it is frequently mixed in road construction. The antistrippant is particularly useful if the rock is wet. It functions by adsorbing onto the surface of the rock, which usually is silica (compare flotation, Part II, Sec. 7.2). Adsorption is so strong that water on the surface of the rock is replaced. The fatty tails point away from the surface of the rock and are solvated by the asphalt. By so doing aminoamides and imidazolines facilitate the bonding between the asphalt and the rock.

3.2.5   AZELAIC AND PELARGONIC ACIDS

Azelaic and pelargonic acids are made from oleic acid (Sec. 3.2.1). A new pelargonic acid process is being developed by Celanese and is described in Part II, Section 10.10.1.

Both azelaic and pelargonic acids are useful for synthesizing specialty polymers and polyesters for synthetic lubricants (Part II, Sec. 10.10.1). Thus a typical synthetic automotive lubricant useful at very low temperatures is the diisodecyl ester of azelaic acid. The odd number carbon dibasic acid is said to be more effective than those with an even number. In comparison with petroleum lubricants it is said to provide greater lubricity, better engine protection, and lower gasoline consumption. Oil-derived hydrocarbons may be blended with azelate esters, and it is likely that they will be used widely in automobiles in the next decade.

Odd number carbon atom acids are rare in nature and in industrial chemistry. Azeleic and pelargonic are the most accessible. It appears that odd number carbon acids are more surface active than are those with an even number.

3.2.6 FATTY ALCOHOLS

Fats and oils may easily be "split" to fatty acids and glycerol (Sec. 3.2.1). They can also be reduced or hydrogenated to fatty alcohols and glycerol (Fig. 3.2). Because the fatty acid groups are mixed, a mixture of fatty alcohols results. The reduction was originally carried out with sodium in ethanol, the Bouveault-Blanc reaction, but hydrogen and a copper chromite catalyst at pressures above 200 atm. are now used. The products differ in that the latter method hydrogenates all the double bonds in the fatty alcohol whereas the Bouveault-Blanc procedure leaves them intact. A newer method using an undisclosed catalyst preserves the double bonds.

In practice it is easier to convert the fatty acid in the triglyceride to its methyl ester by alcoholysis with methanol and then to hydrogenate the methyl ester to the fatty alcohol and methanol. Most vegetable-oil-based alcohols are made in this way.

Fatty alcohols as well as α-olefins may also be obtained from ethylene by use of aluminum trialkyls (Sec. 2.4). Here again petrochemicals have made an impact on traditional processes, but the vegetable-oil-based route is still competitive, at least when vegetable oil prices are low. Over 600 million lb of fatty alcohols were produced by both processes in the United States in 1977.

The special role now occupied by straight chain primary alcohols in detergent technology (Part II, Sec. 7.3.1) results not only from their excellent detergent properties but also because products based on them biodegrade more quickly than compounds containing a benzene ring.

3.2.7   EPOXIDIZED OILS

Unsaturated fats and oils can be epoxidized so that some of the double

bonds are replaced by $-\overset{\overset{\displaystyle O}{\diagup\diagdown}}{C}-C-$ groups. These compounds are added to
poly(vinyl chloride) (PVC) often together with metal soaps to prevent
degradation by light and heat (Part II, Sec. 2.2.1).

Epoxidized oils are also secondary plasticizers, that is they add their
softening power to that of any plasticizer that softens PVC when used on
its own. Epoxidation of fatty acid esters such as butyl or hexyl oleate or
isooctyl "tallate" gives a primary PVC plasticizer (Part II, Chap. 11).
Such materials are widely used in the United States where soy oil is
produced on a large scale. In 1974 in the United States 154 million lb of
epoxidized esters were produced, representing 8% of total plasticizer
production. In Western Europe, on the other hand, epoxidized oils are
normally used only for their stabilizing properties.

3.2.8   RICINOLEIC ACID

Ricinoleic acid, having a hydroxyl group (Fig. 3.1), is the "odd-man-out"
among naturally occurring fatty acids and is found as its triglyceride only
in castor oil. About 120 million lb of castor oil are consumed yearly in the
United States, the largest application being paints and varnishes. Dehydration of the acid gives an isomer of linoleic acid that can be used in
nonyellowing protective coating compositions. Castor oil itself may also
be dehydrated to give a useful drying oil and it may be sulfated to give
Turkey Red oil, a textile dye leveler. It is also used as a polyol (three
hydroxyls per molecule) in polyurethane production.

$$CH_3(CH_2)_5\underset{\underset{\displaystyle OH}{|}}{CH}-CH_2-CH=CH(CH_2)_7C\overset{\displaystyle O}{\underset{\underset{\displaystyle G}{|}}{\diagdown O}} \quad \overset{H_2SO_4}{\longrightarrow}$$

G = Glycerol nucleus

$$CH_3(CH_2)_5\underset{\underset{\displaystyle OSO_3H}{|}}{CH}-CH_2-CH=CH(CH_2)_7C\overset{\displaystyle O}{\underset{\underset{\displaystyle G}{|}}{\diagdown O}}$$

Turkey red oil

Alkaline fusion of ricinoleic acid gives 2-octanol and sodium sebacate.

$$CH_3(CH_2)_5\underset{\underset{OH}{|}}{CH}CH_2CH{=}CH(CH_2)_7COOH \xrightarrow{NaOH} CH_3(CH_2)_5CHOHCH_3$$

2-Octanol

Ricinoleic acid

$+NaOOC(CH_2)_8COONa$

Sodium sebacate

2-Octanol was an important foam depressor prior to the advent of the silicones. Sebacic acid is condensed with hexamethylene diamine to give the specialty polyamide, nylon 6,10. Its dioctyl ester is an excellent PVC plasticizer, but its properties rarely justify its high price.

Thus far castor oil is the only source of sebacic acid, although a Japanese process has been described for the electrodimerization of two molecules of adipic acid with elimination of two molecules of CO_2.

Dry distillation of ricinoleic acid breaks the carbon chain between the eleventh and twelfth carbon atoms and leads to *n*-heptaldehyde and undecyclenic acid. The heptaldehyde may be reduced to *n*-heptanol, which is an acceptable plasticizer alcohol, and the undecylenic acid is treated with HBr, which adds "anti-Markownikoff." Replacement of the bromine with NH_2 gives ω-aminoundecanoic acid (the real reason for the process), which can then be polymerized to nylon 11. Nylon 11 is made only in France, Japan, West Germany, and Brazil. France, in particular, has gone to great lengths to promote the production of castor oil in her former North African colonies. Undecylenic acid, as its zinc salt, is a fungicide effective against *tinea pedis*, or athlete's foot.

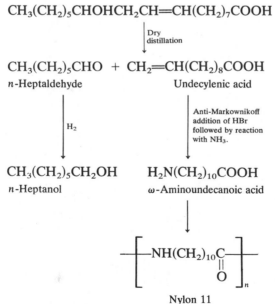

$$CH_3(CH_2)_5CHOHCH_2CH{=}CH(CH_2)_7COOH$$

Dry distillation

$$CH_3(CH_2)_5CHO \; + \; CH_2{=}CH(CH_2)_8COOH$$

n-Heptaldehyde   Undecylenic acid

H_2

Anti-Markownikoff addition of HBr followed by reaction with NH_3.

$$CH_3(CH_2)_5CH_2OH \qquad H_2N(CH_2)_{10}COOH$$

n-Heptanol ω-Aminoundecanoic acid

$$\left[-NH(CH_2)_{10}\underset{\underset{O}{\|}}{C}-\right]_n$$

Nylon 11

3.2.9 GLYCEROL

We have already described the production of glycerol from fats and oils, from propylene by way of allyl chloride and from propylene by way of acrolein (Sec. 2.5.5). Its greatest single use is in drugs and cosmetics, followed by alkyd resins (Part II Sec. 5.6). It is used also in the food and tobacco industries for its moisturizing, lubricating, and softening characteristics. It serves as a plasticizer for cellophane, and dynamite is glyceryl trinitrate (nitroglycerine) absorbed on wood pulp. Glycerol competes with other polyols such as ethylene glycol, pentaerythritol, and sorbitol as a raw material for polyethers for polyurethanes (Sec. 4.4.2). In 1976, 325 million lb of glycerol were produced in the United States, about 60% from petrochemical sources and the remainder from fats and oils.

3.3 CARBOHYDRATES

Carbohydrate sources for chemicals may be subdivided into four main groups: sugars, starch, cellulose, and the so-called carbohydrate gums. In addition there are miscellaneous sources such as the pentosans found in agricultural wastes from which furfural is made. We consider each of these in turn as a source of chemicals. Fermentation processes are mainly carried out on carbohydrate substrates and we therefore include them under this heading.

3.3.1 SUGARS AND FURFURAL

Glucose may be oxidized to gluconic acid, reduced to sorbitol, and esterified to α-methylglucoside as shown in Figure 3.5. Gluconic acid is primarily used as a food additive, and methylglucoside is of some interest in alkyd resins. Sorbitol is also used in alkyd resins and is the starting material for the synthesis of ascorbic acid, vitamin C. It may be condensed with ethylene oxide to give a range of sorbitol-polyethylene oxide emulsifiers that are used, for example, to stabilize food products such as synthetic "whipped cream."

Oat hulls, corn cobs, sugar cane stalks, wood, and many other vegetable wastes contain polymers (pentosans) of pentose sugars such as arabinose. On dehydration with hydrochloric or sulfuric acid, furfural is produced. Furfural is used as a selective solvent in petroleum refining and in the extractive distillation of butadiene from other C_4 hydrocarbons (Sec. 2.1). With phenol it gives phenol-furfuraldehyde resins, which are used to impregnate abrasive wheels and brake linings. Reduction of furfural gives

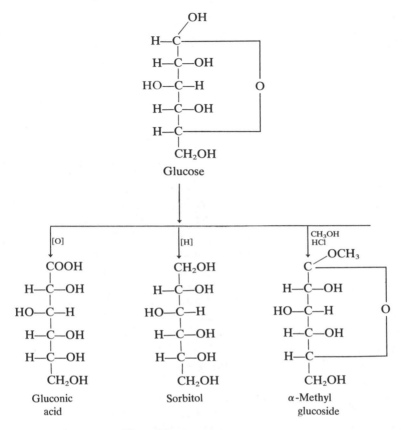

Figure 3.5 Reactions of glucose.

furfuryl alcohol and tetrahydrofurfuryl alcohol, and the latter can be dehydrated and its ring expanded by passage over an alumina catalyst at 270°C. The product is 2,3-dihydropyran. The largest use for furfural in the United States is for conversion to furfuryl alcohol, which in turn undergoes an acid-catalyzed condensation polymerization to yield resins useful as binders in the preparation of foundry cores for molding. Tetrahydrofuran is made by the decarbonylation of furfural with a zinc-chromium-molybdenum catalyst followed by hydrogenation. Tetrahydrofuran may also be made from the condensation of acetylene with formaldehyde (Sec. 2.11), and one of the two US producers uses this process. All of these furan derivatives are useful solvents and chemical intermediates.

$$\underset{\text{Furfural}}{\begin{array}{c}\text{CH}-\text{CH}\\ \| \quad\quad \|\\ \text{CH}\quad\text{C}-\text{CHO}\\ \diagdown\text{O}\diagup\end{array}}\quad\xrightarrow{\text{[H]}}\quad\underset{\substack{\text{Furfuryl}\\ \text{alcohol}}}{\begin{array}{c}\text{CH}-\text{CH}\\ \| \quad\quad \|\\ \text{CH}\quad\text{C}-\text{CH}_2\text{OH}\\ \diagdown\text{O}\diagup\end{array}}\quad+\quad\underset{\substack{\text{Tetrahydrofurfuryl}\\ \text{alcohol}}}{\begin{array}{c}\text{CH}_2-\text{CH}_2\\ | \quad\quad\quad |\\ \text{CH}_2\quad\text{CH}-\text{CH}_2\text{OH}\\ \diagdown\text{O}\diagup\end{array}}$$

$\downarrow \begin{smallmatrix}\text{Al}_2\text{O}_3\\ 270^\circ\text{C}\end{smallmatrix}$

2,3-Dihydropyran

A specialty nylon, nylon 4, which results from the polymerization of 2-pyrrolidone, is related to furfural chemistry since the monomer can be synthesized from tetrahydrofuran. Nylon 4 has the interesting property of "moisture regain" which makes it more like cotton than the other nylons. Thus far however it has made no commercial headway.

There is only one producer of furfural in the United States, Quaker Oats Company, and production is estimated at less than 200 million lb. Extensive growth is not expected. Foster Wheeler is building a furfural plant in Kenya that will process corn cobs to give 5000 tonnes year^{-1} furfural, 3300 tonnes year^{-1} acetic acid, and 3000 tonnes year^{-1} formic acid. As about 6 tons of corn cobs are required for every ton of liquid product the collection of the feedstock will present problems even in a country where labor is cheap.

Another recent development that bears on furfural is the interest in the sugar alcohol, xylitol, as an ingredient in chewing gum, candy, and sweet cereals to prevent dental caries. A plant in Finland was producing xylitol, and another plant was planned for the United States. Xylitol like furfural is derived from pentosans. It has been removed from the market, however, because tests showed it was potentially carcinogenic.

$$n\text{-C}_5\text{H}_{10}\text{O}_5 \xrightarrow[\text{H}_2\text{O}]{\text{H}_2\text{SO}_4} n\text{-HO}-\text{CH}_2-\text{CHOH}-\text{CHOH}-\text{CHO}$$

Pentosan   Pentose

$$\underset{\text{Furfural}}{\begin{array}{c}\text{CH}-\text{CH}\\ \| \quad\quad \|\\ \text{CH}\quad\text{C}-\text{CHO}\\ \diagdown\text{O}\diagup\end{array}}\qquad\qquad\underset{\text{Xylitol}}{\begin{array}{c}\text{CH}_2\text{OH}\\ |\\ \text{H}-\text{C}-\text{OH}\\ |\\ \text{HO}-\text{C}-\text{H}\\ |\\ \text{H}-\text{C}-\text{OH}\\ |\\ \text{CH}_2\text{OH}\end{array}}$$

3.3.2 STARCH

Starch is one of the most important chemical products of the vegetable kingdom and is found in practically all plant tissues, especially in seeds (e.g., wheat and rice) and tubers (e.g., potatoes). Commercial starch in the United States is extracted chiefly from corn and to a lesser degree from wheat. In Europe potatoes are an important source. Chemically, it comprises two distinct polymers of α-D-glucopyranoside.

The linear polymer, amylose, is made up of several hundred glucose units connected by α-D-$(1 \rightarrow 4)$ glucosidic linkages (**a**, Fig. 3.6). The branched chain polymer, amylopectin (**b**, Fig. 3.6), is of much higher molecular weight and contains between 10,000 and 100,000 glucose units. The segments between the branched points contain about 25 glucose units joined as in amylose while the branched points are linked by α-D-$(1 \rightarrow 6)$ bonds. Most cereal starches contain about 75% amylopectin and 25% amylose.

Because amylose has straight chains, its molecules can approach each other closely and form hydrogen bonds. This interaction is so strong that amylose can scarcely be dispersed in water. Amylopectin, on the other hand, has branched chains that have more difficulty in approaching each other, and it is therefore readily dispersed.

If starch, which is an amylose/amylopectin mixture, is dispersed in water, a gel results that forms a "skin" on standing. The skin is a result of hydrogen bonding. So-called waxy starches such as waxy maize starch, which arose from genetic experimentation, are very high in amylopectin, and consequently they form gels that do not "skin."

Amylose is prevented from skinning if it is converted into a derivative such as the phosphate. Only a small number of phosphate groups per molecule are required to interfere with the hydrogen bonding. Consequently starch phosphates are widely used as thickeners in the food industry.

Starch is used in adhesives and as a size in textiles and paper manufacture. Sizing is a method for altering the surface properties of paper fibers. For example, many grades of paper are coated with an aqueous suspension of pigments (such as clay) in adhesives (such as starch) to provide a smoother surface, control the penetration of inks, and generally improve the paper's appearance. Dextrinized or degraded starch is used as an adhesive.

Starch may be modified chemically. About half the starch produced in the United States is hydrolyzed with hydrochloric acid to glucose or partially hydrolyzed starch-glucose syrups that can be sold as such or isomerized to fructose (Sec. 3.4.1) by way of a new process. Hydroxyethylstarch, made by the action of ethylene oxide on starch, is used in

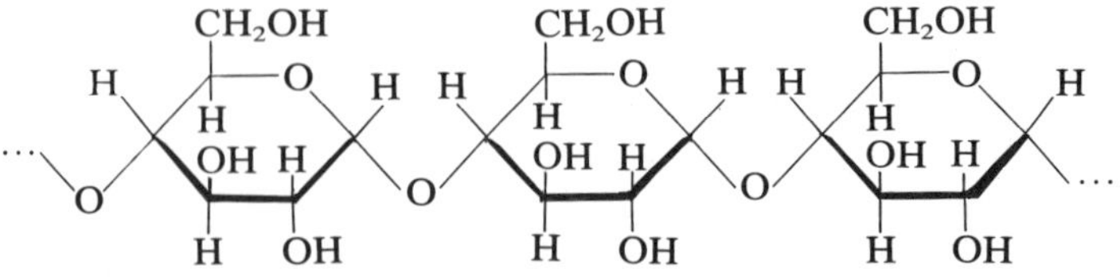

Part of amylose molecule showing 1–4 α-glucosidic linkages

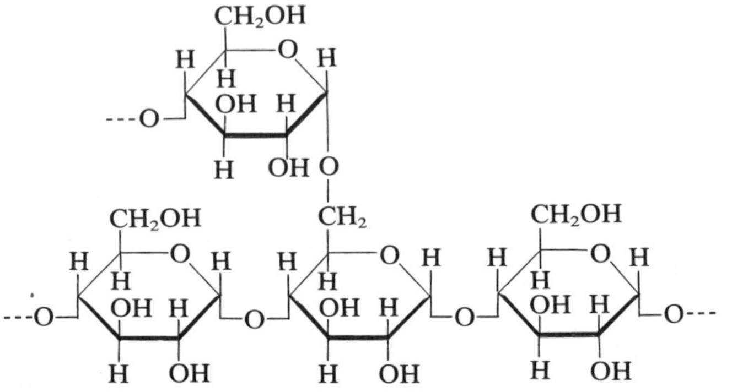

Part of amylopectin molecule showing chain branching and 1–6 bonds

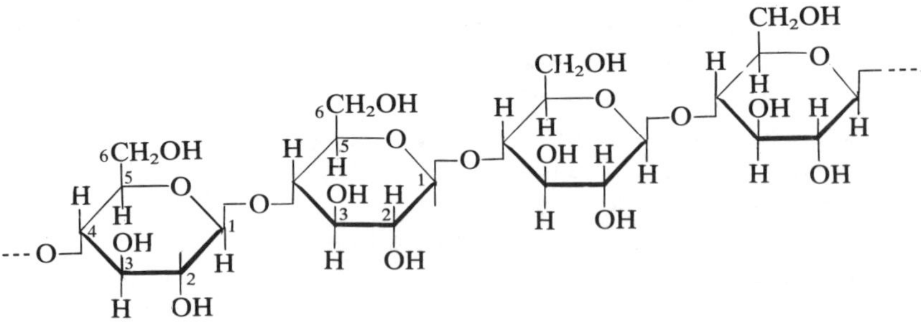

Part of cellulose molecule showing 1–4 β-glucosidic linkages

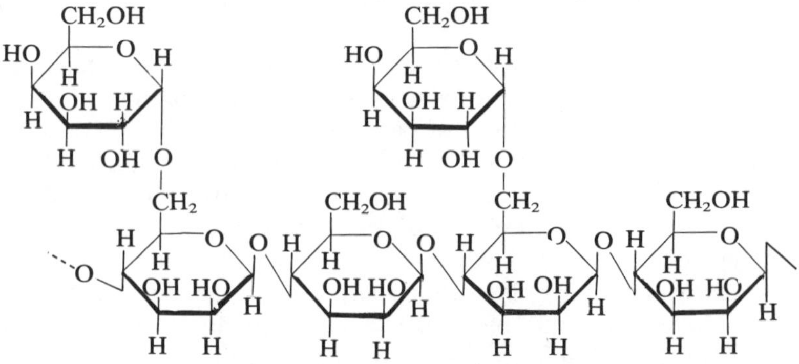

Part of guar gum molecule showing chain of mannose units with pendant galactose units

Figure 3.6 Starch, cellulose, and guar structures.

paper coating and sizing because it disperses more easily than starch itself and provides a dispersion of better clarity. Starch may also be reacted with cationic reagents such as dimethylaminoethyl chloride to give "cationic starch," which is used to impart greater strength to paper products. Acrylonitrile may be polymerized onto a starch polymer with the aid of a ceric sulfate catalyst to give a graft copolymer capable of absorbing large amounts of water.

3.3.3 CELLULOSE

Cellulose is the primary substance of which the walls of vegetable cells are constructed. It occurs in plants, wood, and natural fibers, usually combined with other substances such as lignin, hemicellulose, pectin, fatty acids, and rosin. It accounts for about 30% of all vegetable matter. It may be represented by the formula $(C_6H_{10}O_5)_n$.

Cellulose is a linear polymer composed largely of glucose residues in the form of anhydroglucopyranose joined together by β-glucosidic linkages (c, Fig. 3.6). The β-linkages make the cellulose molecule very stiff compared with amylose, which has α-glucosidic linkages. It also makes cellulose more difficult to hydrolyze than starch. It is this difficulty that makes cellulose indigestible by humans. Accordingly, one of its uses is in diet foods, providing bulk and satiety but not calories. Had nature not insisted on β-linkages, a vast source of nutrition would be available to humans. On the other hand, cellulose with α-linkages would not be stiff and would not possess the structural properties it has in wood or the strength it has in textiles. Cows and other ruminants have digestive systems containing enzymes that hydrolyze cellulose, and they are therefore able to use grass and similar materials for food.

Cellulose may be obtained from wood or derived in very high purity from cotton linters. The majority of production is used in paper manufacture. In the manufacture of pulp used in cheap paper the wood fibers are simply pulped mechanically. In the premium chemical pulp lignins, fatty acids, and rosins are first removed by sodium hydroxide treatment.

Cellulose also has a range of chemical uses. Cellulose derivatives (Fig. 3.7) may be water-soluble like methylcellulose, hydroxyethylcellulose, or sodium carboxymethylcellulose, or water-insoluble like the cellulose esters and ethylcellulose.

Methylcellulose, carboxymethylcellulose, and hydroxyethylcellulose are all thickening agents and form protective colloids. For example they are used in water-based latex paints to produce desirable flow and viscosity and to help stabilize the emulsion and pigment dispersions. Protective colloids affect the washability, brushability, rheological properties, and

CELL—O—CH$_3$ Methylcellulose

CELL—O—C$_2$H$_4$OH   Hydroxyethylcellulose

CELL—O—CH$_2$—COONa   Sodium carboxymethylcellulose

$$\text{CELL—O—C}_2\text{H}_4\text{—N} \begin{cases} \text{C}_2\text{H}_5 \\ \text{C}_2\text{H}_5 \end{cases}$$ Diethylaminoethylcellulose

Water-soluble derivatives

CELL—O—COCH$_3$ Cellulose acetate

CELL—O—NO$_2$ Cellulose nitrate

CELL—O—C$_2$H$_5$ Ethylcellulose

CELL—O—COC$_2$H$_5$ Cellulose propionate

Water-insoluble derivatives

Figure 3.7   Cellulose derivatives. (Cellulose is represented as CELL-OH).

color acceptance of a paint. The same properties make them useful in foods such as ice cream and in inks and adhesives.

Methylcellulose is also used as a base for paper coatings since it is a good film former. Hydroxyethylcellulose is an adhesive and a binder in woven fabrics. Sodium carboxymethylcellulose (CMC) has a major use as a soil suspending agent in detergents, for if it were not present the dirt would tend to redeposit on the articles being washed. Instead the CMC forms a protective colloid and holds it in suspension. CMC is used in textile sizing, paper coating, and in oil well drilling mud, a material that helps to bring to the surface the dirt and rock particles dislodged by the drill. Diethylaminoethylcellulose is a cationic material useful in cotton finishing.

Cellulose triacetate is made by the action of equal quantities of acetic anhydride and glacial acetic acid on "chemical cotton," a form of cellulose obtained by purification and conversion of cotton linters. All three hydroxyl groups in each glucose unit are acetylated. Cellulose acetate is obtained by partial hydrolysis of cellulose triacetate with water and a small amount of acetic and sulfuric acids. Addition of a large excess of water at the appropriate stage stops the hydrolysis and precipitates the acetate (with an average of two acetyls per glucose unit) as flakes.

Cellulose acetate and triacetate may be used as plastics molding materials or spun into fibers and used for textiles. Cellulose triacetate is more difficult to process than the acetate but can be heat-set to give wash-and-wear fabrics. Cellulose propionate, acetate-propionate, and acetate-butyrate are also used as plastics materials and for films and lacquers. Ethylcellulose is useful in coatings and some plastics applications.

$$
\left[\begin{array}{c} CH_2OH \\ \text{(glucose ring)} \\ OH \ H \\ H \ \ OH \end{array} \right]_n + 3n \ \begin{array}{c} CH_3C \diagup^{O} \\ \diagdown O \\ CH_3C \diagup \diagdown_O \end{array} \xrightarrow{H_2SO_4} \left[\begin{array}{c} CH_2O.OC.CH_3 \\ \text{(glucose ring)} \\ O.OC.CH_3 \ H \\ H \ \ O.OC.CH_3 \end{array} \right]_n
$$

Cellulose   Cellulose triacetate

Cellulose nitrate or nitrocellulose was an early explosive (guncotton), plastic (celluloid), and surface coating (lacquer). It was the first material used for coating automobiles by assembly-line procedures, its rapid drying making semiautomated production possible. It is obtained by nitration of cellulose, about two nitro groups per glucose unit being introduced. It is also used for fabric coating, as the basis for lacquers for furniture (where it is being displaced by "melamine precatalyst" formulations) (Part II, Sec. 2.3.1), and in plastic moldings and film.

In addition to chemical conversion, cellulose may be altered physically or "regenerated." Two processes are used, both of which start with highly purified cellulose from wood pulp, and the products are known as viscose rayon and cuprammonium rayon. Viscose rayon is produced by conversion into the soluble xanthate by "ripening" with concentrated sodium hydroxide followed by treatment with carbon disulfide:

$$
-\overset{|}{\underset{|}{C}}-OH \xrightarrow{NaOH} -\overset{|}{\underset{|}{C}}-ONa \xrightarrow{CS_2} -\overset{|}{\underset{|}{C}}-O-\overset{}{\underset{\underset{S}{\|}}{C}}-SNa
$$

Cellulose   Cellulose xanthate

0.5–0.6 xanthate groups per glucose unit are inserted. The solution is known as "viscose." It is extruded through a spinerette—a metal disc with many tiny holes— into an acid coagulating bath that regenerates the cellulose as a fiber. The fibers are washed and spun, that is oriented by stretching. The carbon disulfide is also regenerated and is recycled. The xanthation serves simply to solubilize the cellulose in a form from which it can readily be regenerated.

The premium cuprammonium rayon is made by dissolution of cotton linters or wood pulp in ammoniacal copper oxide. The solution is extruded through spinnerettes into an acid bath in the same way as viscose.

Cuprammonium rayon is chemically similar to viscose rayon but gives a finer yarn that is used in sheer fabrics.

Cellophane film is also made from viscose solution, but the cellulose is regenerated in sheet form instead of fibers. It has a high moisture transmission and is lacquered with a waterproofing agent, traditionally cellulose nitrate, which also imparts heat sealing capability. Polyvinylidene chloride (Part II, Sec. 2.1.7) is now widely used for this purpose. It is still a major packaging film despite the advent of plastic films such as polyethylene and Saran.

Cellulose sponges are also made from xanthate, which is cast into blocks together with sodium sulfate crystals of various sizes. The xanthate is decomposed to regenerate cellulose, and when the sulfate is washed out it leaves holes in the sponge. Recovery of sodium sulfate is a major part of the operation. Of course, in all processes involving cellulose xanthate the carbon disulfide is recovered and recycled.

The strength properties of cellulose, and to a degree its ability to hydrogen bond, make it useful for the formation of paper and nonwoven fabrics. For paper, cellulose is "beaten" or "pulped" until very finely divided particles result. Glassine is prepared from very finely divided particles and is translucent. Opaque papers result from less finely divided particles. The wet and dry strength, bulk density, water and oil resistance, and gas permeability of paper are properties that can be modified either by adding chemicals to the pulp or by coating the finished sheet. Certain polymers (Sec. 3.3.2) increase wet strength. Starch and rosin soap (Sec. 3.2.1) are widely used for sizing.

Nonwoven fabrics result when rayon (Sec. 3.3.3) is chopped into very fine particles known as fibrils. These are described in Part II, Section 3.10.

3.3.4 GUMS

Gums, like starch and cellulose, are carbohydrate polymers. They differ from them in that the monomeric unit may be a sugar other than glucose, and the chemical configuration and the way in which the units are joined may be different.

The molecular weight of gums is usually between 200,000 and 300,000, that is about 1500 monomer units. Guar (**d**, Fig. 3.6) is a typical gum. It consists of a chain of mannose units joined by 1,4-glycosidic linkages, and attached to every other mannose unit is a pendant galactose unit.

The main gums and their origins are shown in Table 3.1. Each gum has characteristic properties slightly different from other gums. Frequently the commercially important differences lie in the rheological properties of the dispersions of the gums in water.

Table 3.1 Natural Gums

Source	Examples
Plant seeds	Guar gum, locust bean gum
Seaweed extracts	Alginates, carrageenan, agar
Tree exudates	Gum arabic, karaya gum, gum tragacanth
Citrus fruits	Pectin
Animal skin and bones	Gelatin
Fermentation	Xanthan gum

Gums, like cellulose, may be chemically modified, and the most useful derivatives are carboxymethyl, hydroxypropyl, and dimethylaminoethyl gums.

The applications of gums are wide, but guar gum may be considered typical. It is used in paper manufacture since it helps to retain the very fine cellulose particles in the paper matrix thus increasing the yield of product. Also, since fewer of these "fines" are in the water that drains from the machine, the pollution problem is diminished. At the same time, guar strengthens the paper by hydrogen bonding to the fibers and helping them achieve a linear rather than a random configuration.

Guar is also useful as a flocculent for precipitating mineral slimes and as a suspending agent for ammonium nitrate that not only leads to a much cheaper explosive than dynamite or nitroglycerine but also to one that is more effective because it assumes the shape of the cavity where the blast is to start. Guar has many times the thickening power of starches and may be used in combination with them.

Carboxymethylguar gum is an anionic material useful as a print gum paste. This means that it serves as a binder for a pigment used to impart color and design to cloth. In contrast, diethylaminoethylguar gum is cationic and is a particularly effective flocculent of "fines," tiny bits of particulate matter onto which it adsorbs.

An interesting gum only recently commercialized is Xanthan gum, obtained not from animals or plants but by fermentation of carbohydrates with a bacterium, *Xanthamonus campestris*. The gum is a complex glucose polymer, and its aqueous solutions are unusually stable—showing unchanged viscosity over broad temperature, salt concentration, and pH ranges. The product is therefore used to thicken oven cleaners based on strong alkali as well as the acid solutions used as metal cleaners. Its largest use is as a component of oil well drilling mud that must contain saline water. Its largest potential use is in so-called tertiary oil recovery where it thickens the water used to "push" oil, unobtainable in any other way, through the dense, oil-bearing rock formation. Unlike other gums such as guar, Xanthan does not adhere to surfaces. If it did, it would deposit as a film on the rock surface and lose its thickening power.

3.4 FERMENTATION

When supplied with suitable nutrients, single cell microorganisms—yeasts, molds, fungi, algae, bacteria including the very important antibiotic-producing actinomyces—thrive and multiply. As they do so various waste products of their metabolisms accumulate. The microorganisms can tolerate only low concentrations of their own wastes. Nonetheless, under certain circumstances these wastes, which can be either intra- or extracellular, can be concentrated and used. The process is known as microbiological conversion, or fermentation, and it occurs as a result of the catalytic action of various enzymes produced by the microorganisms on the nutrient or substrate. Thus fermentation can also be brought about by pure enzymes or portions of cells that contain enzymes such as mitochondria.

The substrate is usually but not invariably a carbohydrate. There has been interest recently in the use of gas oil or other petroleum hydrocarbon substrates for the production of single cell protein. Much of the development work was done in the days of abundant petroleum, and cynics are already asking if the reverse process—petroleum from single cell protein—is also operable. Following political troubles in Italy, BP is dismantling its gas-oil-based plant in Sardinia, and its pilot plant in Scotland is also being mothballed. An Imperial Chemical Industries plant based on methanol (which could presumably be obtained from coal) survives, and a full-scale plant is promised for 1980. The product is regarded as suitable for animal feed but not yet for human consumption.

In the United States inexpensive soy protein is available, and there is little reason to believe that its production will not prosper in the future. Soybean production is also being developed on a large scale in Brazil, and there should be adequate supplies of vegetable protein, particularly for animal feed, for the foreseeable future, at least in the Western Hemisphere. On the other hand, there are many protein-short countries in the world where arable land is preempted for food production for humans. There the production of protein by fermentation may be helpful.

At present, fermentation is customarily used in the chemical industry only when an economic chemical process is not available. Its largest volume application is in sewage treatment where obnoxious amines and sulfur compounds are aerobically oxidized to nitrates and sulfates or anaerobically digested to methane and carbon dioxide. The next largest application is the production of alcoholic beverages where the product can either be consumed in its dilute form (beer and wine) or concentrated by distillation (whisky, brandy, gin, vodka).

The fermentation reaction itself is usually efficient and inexpensive, and the conditions required are mild. "Waste heat" from power stations, refineries, and factories could easily be used as an energy source. On the other hand, nutrients tend to be expensive, reactions are slow, and the product is dilute, so that huge tank capacities are required. Unless the product precipitates (e.g., single cell protein), its isolation can be tedious and expensive. If the fermentation is aerobic, mass transfer of oxygen to the required site demands intricate engineering. Nonetheless, there are products for which fermentation methods are uniquely suitable. The main groups of antibiotics—penicillins, streptomycins, and tetracyclines—are all made by fermentation. These compounds are discussed in Part II, Section 8.7. Penicillin has been synthesized in the laboratory, but chemical processes are used commercially only for modification of the basic structure.

The synthesis of L-amino acids by fermentation has been pioneered largely in Japan. Production of every essential amino acid by fermentation is now possible. Because demand is small many of them are still made by chemical methods followed by resolution of the resulting DL-compound. The most important amino acids made by fermentation are L-glutamic acid, whose sodium salt is a food flavoring agent and flavor enhancer, and L-lysine, which is a nutritional supplement for animal feeds and grains deficient in it.

A dramatic but small volume application made possible by the availability of high purity amino acids is intravenous feeding or hyperalimentation. The patient is "fed" amino acid solutions of high concentration through the jugular vein, sometimes for months at a time. The technique is especially valuable for infants suffering from metabolic disturbances from which they might not otherwise recover.

A classic example of the use of fermentation to supplement the chemist's synthetic capabilities is found in the synthesis of vitamin C (ascorbic acid) shown in Figure 3.8. Synthesis of L-sorbose without fermentation would present a formidable problem in organic chemistry. The appropriate bacterium, however, *Acetobacter suboxydans*, selectively oxidizes the hydroxyl group on the C_2 of sorbitol, making feasible the remainder of the synthesis.

The 11-hydroxylation of progesterone in the synthesis of cortisone is another example of the power of fermentation (Part II, Sec. 8.8). Without it cortisone therapy could never have developed.

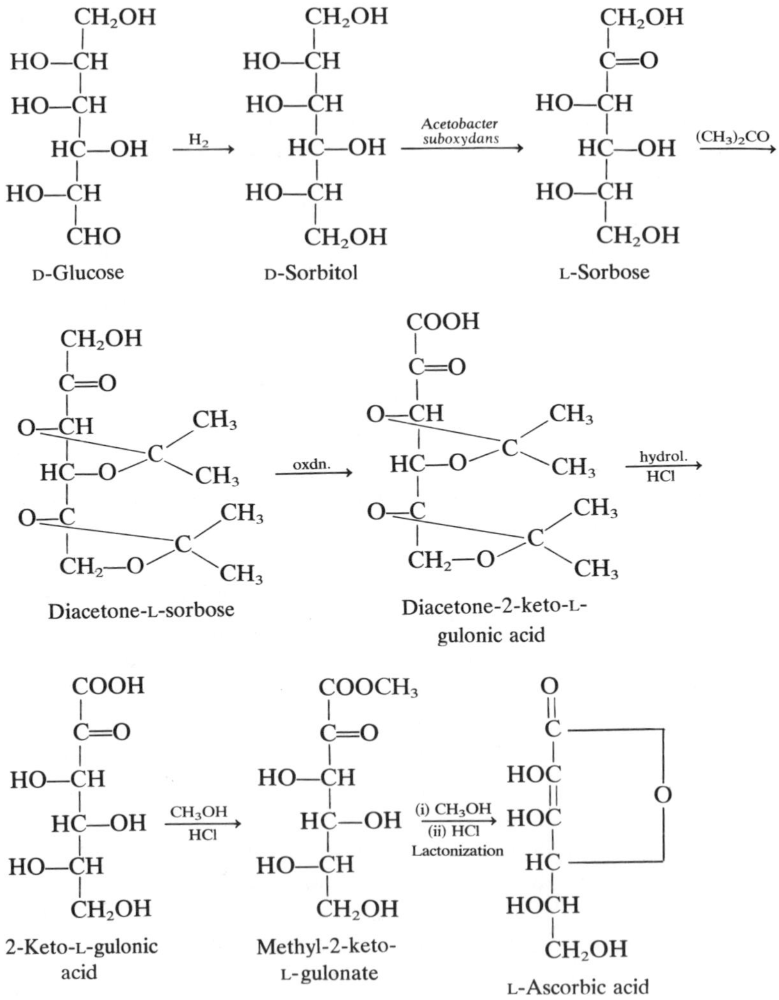

D-Glucose $\xrightarrow{H_2}$ D-Sorbitol $\xrightarrow[\text{suboxydans}]{\text{Acetobacter}}$ L-Sorbose $\xrightarrow{(CH_3)_2CO}$

Diacetone-L-sorbose $\xrightarrow{\text{oxdn.}}$ Diacetone-2-keto-L-gulonic acid $\xrightarrow[\text{HCl}]{\text{hydrol.}}$

2-Keto-L-gulonic acid $\xrightarrow[\text{HCl}]{CH_3OH}$ Methyl-2-keto-L-gulonate $\xrightarrow[\text{Lactonization}]{\text{(i) } CH_3OH \text{ (ii) HCl}}$ L-Ascorbic acid

Figure 3.8   Synthesis of vitamin C.

Citric and lactic acids are both made from sucrose wastes by fermentation. Citric acid may also be made from hydrocarbon substrates. Tartaric acid could be similarly manufactured, but extraction from wine industry residues is easier.

3.4.1   ENZYMES

A major area of fermentation research is the enzymes themselves. These can be produced by fermentation and can digest many of the wastes that contribute to pollution. They have also been incorporated into detergent formulations where they are intended to hydrolyze protein-based stains such as blood (Part II, Sec. 7.6.8).

One problem of enzyme technology was solved when it was realized that enzymes retain their activity when attached to an insoluble matrix by a chemical linking agent such as glyoxal, OHC.CHO. In all likelihood the enzyme is not truly "fixed" and retains the ability to accommodate itself to the shape and conformation of the molecules whose chemical reactions it is catalyzing. This "immobilization" means that enzymes can be reused many times and also that continuous rather than batch reactors are possible.

One of the first applications for immobilized enzymes in the United States has been the conversion of starch to glucose followed by isomerization of the glucose to fructose. Fructose is sweeter than glucose, and its syrups are useful for manufacture of soft drinks, confectionery, and sweet goods generally. The reaction requires three enzymes. The first is an amylase that degrades the starch to lower molecular weight polymers; the second is an amyloglucosidase that converts these oligomers to glucose; and the third is an isomerase that changes the glucose to fructose. Attempts to introduce the process to Europe have resulted in protests from local sugar producers, and it is possible that tariff barriers will render it uneconomical there. Large capacity, on the other hand, has been installed in the United States largely as a response to an unprecedented increase in sucrose prices in the early 1970s.

DL-Amino acid mixtures may also be resolved by use of immobilized enzymes. If a racemic mixture is acetylated and passed over an immobilized L-acylase, the acetyl group is hydrolyzed from the L-amino acid, but the D-amino acid remains unchanged. The mixture may then be separated by crystallization.

3.4.2 A FERMENTATION SCENARIO

In the period between the world wars fermentation was the most popular route to aliphatic organic chemicals. Production was expensive; tonnages were low. Nevertheless, reaction pathways existed by which most present-day organic chemicals were made, and with modern technology they could probably be made more cheaply and efficiently than they were then. Thus modern technology makes possible the synthesis of the important olefins—ethylene, propylene, and the butylenes—from fermentation ethanol. Dehydration of ethanol yields ethylene which in turn may be dimerized (Sec. 2.4) to 2-butene. This is a recently discovered reaction believed to have commercial potential. Mixed metathesis (Sec. 2.5.7) between 2-butene and ethylene should yield propylene.

$$CH_3CH_2OH \xrightarrow{-H_2O} CH_2{=}CH_2$$

Ethanol   Ethylene

$$CH_2{=}CH_2 \xrightarrow{\text{catalyst}} CH_3{-}CH{=}CH{-}CH_3$$

2-Butene

$$CH_3{-}CH{=}CH{-}CH_3 + CH_2{=}CH_2 \xrightarrow{\text{catalyst}} 2CH_2{=}CH{-}CH_3$$

Propylene

An alternative approach to the C_4 unsaturates involves the reaction of acetaldehyde with ethanol to yield butadiene.

While most carbohydrates can be hydrolyzed to sugars and fermented to alcohols, the preferred material is sugar cane. It can be cultivated in most warm countries and grows fast. In some areas it is a year-round crop. It harnesses the sun's energy with a relatively high efficiency—about 8% compared with 1% for the best solar cells. The ripe plant, which contains 11–15% sugar, is crushed between rollers to remove the juice, and the crushed cane is further extracted with hot water. Further purification and concentration gives the pure cane sugar plus molasses or, alternatively, the solution can be used as a fermentation feedstock. The residue, known as bagasse, is a cellulose combined with pentosans, lignin, and inorganic salts. Conventionally it is burned to provide heat for processing of the sugar, 1 ton of bagasse at 51% solids being equivalent to one barrel of oil. Alternatively it may be digested with acid to give furfural (Sec. 3.3.1). The bagasse fibers may also be bonded with urea-formaldehyde or phenol-formaldehyde resins and compressed to give chipboard and acoustical wallboard.

The grave drawback to fermentation and the use of biochemical enzymes in a world of expensive energy is that the products are produced in a dilute form, and the cost of removal of water is high because of its high latent heat of evaporation. Heat costs can be reduced if the steam from one evaporator is used to heat another, and the steam from that used to heat another, and so on, but the capital cost of such multistage evaporators to some extent offsets the savings achieved by them. Furthermore, if all the corn, wheat, and other crops grown by world farmers were converted to ethanol, it would still only give 6–7% of the energy equivalent of present world oil production.

As already noted, methane can be produced by anaerobic fermentation of sewage sludge or, indeed, of organic wastes generally. These are referred to as biomass. About 75% of the caloric value of sludge can be recovered in this way, which sounds impressive but in fact serves mainly to make sewage plants independent of outside supplies of power.

Nonetheless, there is interest in broadening the scope of the process. In California, for example, the late Howard Hughes "mined" the garbage dumps for methane, and in southwest England a well-known eccentric runs his car on methane from the wastes produced by his chickens.

Trivial though this may sound, it is estimated that biomass, which includes residues of the forest industry, corn cobs, oat hulls, and various plants, could supply $5-10 \times 10^{15}$ Btu of fuel and chemicals by the year 2020. This would amount to 5–10% of present US energy consumption. The processes nearest to commercialization are

1   Combustion of wood and low moisture plants to produce steam and cogenerated electricity,

2   Gasification of wood and low moisture plants to low Btu gas, intermediate Btu gas (IBG), SNG, and ammonia,

3   Pyrolysis of wood and low moisture plants to give SNG, fuel oil, and char,

4   Anaerobic digestion of manure and high moisture crops to give IBG and SNG.

There is a clear implication that many forms of biomass, especially wood, are better burned than fermented and at present the cheapest way to make chemicals from wood is to make them from the petroleum "saved" by burning wood. This situation may not continue and details of chemicals from wood will be found in notes to Section 3.3.3.

In summary, fermentation is extremely valuable to do what the chemist cannot do (e.g., antibiotic and L-amino acid production, steps in vitamin C and cortisone synthesis). It holds promise of providing a new route to energy (methane synthesis) and to animal and even human food (fermentation of various substrates to protein). Readily produced fermentation ethanol can be dehydrated to the all-important chemical, ethylene, on which a lion's share of our petrochemical industry is based. How much energy, food, or ethylene will be produced by fermentation, however, will depend on economics, which in turn are a function of the seriousness of shortages and the alternative routes devised to alleviate them.

REFERENCES AND NOTES

Sec. 3.1   Producer gas, water gas, and synthesis gas are often confused. Producer gas was a low Btu gas made from air, steam, and coke and is now obsolete. Water gas is the gas obtained from the passage of steam over coke (Sec. 3.1.2) when its composition is typically 40% CO, 50% H_2, 5% CO_2, and 5% N_2 and CH_4. A gas with a higher H_2:CO ratio is obtained by steam reforming of methane and hydrocarbons (Sec. 2.12). For the manufacture of chemicals, a synthesis gas must be produced with the appropriate H_2:CO ratio, and to the extent that this is necessary, water gas is not the same as synthesis gas. Manufacture of ammonia requires a synthesis gas with hydrogen but no carbon monoxide. Methanol and the Fischer–Tropsch reaction require H_2:CO = 2:1 and the oxo reaction requires H_2:CO = 1:1.

The huge classic, H. H. Lowry, Ed., *The Chemistry of Coal Utilization*, originally published in 1945 was reprinted by Wiley, New York, in 1977. A review of modern coal tar technology has been provided by H. C. Messman, *CHEM-TECH*, October 1975, p. 618. Articles on the use of coal as a source of oil in South Africa have been published in *Chem. Week* September 25 1974, p. 33 and in *CHEMTECH*, January 1976, p. 5. Possible chemicals from coal have been described in an article in *Chem. Week* June 12 1974, p. 11. Coal gasification has been discussed in *Mat. Eng.*, July 1974, p. 16. The methanation route from coal to SNG is discussed by D. Rooke, *Chem. Eng.* January 1978, p. 34. A new technique for hydrogenation of coal has been described in *Science*, September 1975, p. 5. How to remove sulfur from coal has been described in *Chem. Week*, January 7 1976, p. 33. The underground gasification of coal is reviewed by R. M. Nadkarny, Bliss, and Watson, *CHEMTECH*, April 1974, p. 230. Coal conversion technology has been described by H. Perry, *Chem. Eng.*, July 22 1974 p. 88 and in an article in *Chem. Eng. News*, December 1, 1975, p. 24.

Processes for making synthetic crude oil from shale and coal are detailed by D. L. Class, *CHEMTECH*, August 1975, p. 499. Coal gasification processes have been reviewed in *Chem. & Eng. News*, November 1, 1976, p. 16. Transportation of coal slurry in pipelines has been described in *Chem. Eng. News*, June 27, 1977, p. 20. The economics of *in situ* coal recovery has been discussed by Goddin, Hall, and Mason, *Hydrocarbon Proc.*, **55** (1976) 109. Coal liquefaction has been described by J. B. O'Hara, *Hydrocarbon Proc.*, **55** (1976), 221. The catalytic hydrogenation of coal to provide gasoline is reviewed in *Chem. Week*, December 21, 1977, p. 37. These articles are representative of the hundreds that have been published on new techniques for coal utilization.

Sec. 3.1.2   According to *Euro. Chem. News*, August 21, 1978, coal is available in South Africa at $5–8 per ton compared with $8–30 in the United States and $65–80 in Germany. A major problem to an extensive coal-based chemical industry would be the amount of coal needed. Bayer has estimated that to produce all the organic chemicals currently made from petroleum would require an extra 250 million tons of coal per year in Western Europe. Current West European production is 275 million tons, and a major expansion would require deep mines.

Sec. 3.2   Information and statistics about fatty acids can be obtained from the Fatty Acid Producers' Council, 475 Park Ave. So., New York, NY 10016. A classical book on fatty acids is T. P. Hilditch, *The Industrial Chemistry of Fats and Waxes*, Baillière, Tindall & Cox, London, 1949. Another important volume is A. W. Ralston, *Fatty Acids and Their Derivatives*, Wiley, New York, 1948. K. S. Markley, Ed., *Fatty Acids: Their Chemistry, Properties, Production, and Uses*, 5 vols., 2nd ed., Interscience, New York, 1960–1968, is more recent and more extensive than the two preceding volumes.

Reviews involving fatty acid technology are found in J. K. Craver and R. W. Tess, *Applied Polymer Science*, American Chemical Society, Washington, DC, 1975, Chapters 36 and 51.

Sec. 3.2.2 Fatty nitrogen chemistry is reviewed in E. S. Patterson, Ed., *Fatty Acids and Their Industrial Applications*, Dekker, New York, 1968, Chapter 5.

Sec. 3.3.3 Cellulose makes up about 50% of wood together with 25% of lignin (phenyl-propane polymers) and 25% of hemicelluloses (carbohydrate polymers built up from molecules of simple sugars). Wood is thus a source of such polymers. In prehistoric times, however, it was also a precursor of coal and oil. In principle, therefore, it might be possible to obtain petrochemicals from wood. Given present technology, hydrolysis of cellulose to sugars followed by conversion to chemicals such as ethanol, furfural, and lactic acid is potentially attractive. Lignin can be hydrogenated to phenol. In the absence of coal, wood can be gasified to ammonia, methanol, and hydrocarbons. The use of wood as a feedstock is discussed by I. S. Goldstein, *Science*, **189** (1975) 847; *A. I. Chem. E. Symposium Series*, **74** (1978) 111; and in a paper "Chemicals from Wood: Outlook for the Future" to the 8th World Forestry Congress, Jakarta, Indonesia, October 1978.

Sec. 3.3.4 An excellent book on carbohydrate gums' chemistry and technology is R. L. Whistler and J. N. BeMiller, Eds., *Industrial Gums: Polysaccharides and Their Derivatives*, 2nd ed., Academic, New York, 1973.

Sec. 3.4 The production of single cell protein is described in N. Calder, "Food from Gas Oil," *New Sci.*, **36** (1967), 468 and in publications from BP Ltd., Britannic House, Moor Lane, London EC2. Protein from methanol together with a list of other substrates appears in *New Protein*, ICI Educational Publications, Mill-bank, London SW1P 4QG.

A review of the fermentation industry has been published by D. Pearlman, *CHEMTECH*, July 1977, p. 434.

An overview of single cell protein technology has been provided by D. D. MacLaren, *CHEMTECH*, October 1975, p. 594. ICI's plans to build a large plant for production of single cell protein is described in *Eur. Chem. News*, October 1, 1976, p. 4.

High fructose corn syrup technology has been described in *Chem. Week*, October 29, 1975, p. 33. Fermentation of biomass to provide methane and methanol is described in *Can. Chem. Proc.*, August 1977, p. 76.

For information on continuous fermentation see A. C. R. Dean, D. C. Elwood, C. G. T. Evans, and J. Melling, *Continuous Culture*, Ellis Horwood, 1977.

Sec. 3.4.1 An excellent article on the possibilities of using enzymes in nitrogen fixation has been written by K. J. Skinner, *Chem. Eng. News*, October 4, 1976, p. 22.

Sec. 3.4.2 The question of chemicals and energy from biomass is tackled by I. S. Goldstein, cited above, and was discussed at an American Chemical Society meeting, April 1979, reported in *Chem. Age*, May 5, 1979, p. 294.

Chapter Four
HOW POLYMERS ARE MADE

The polymer industry stands out above all others as a consumer of heavy organic chemicals, and it converts these to the products we call plastics, fibers, elastomers, adhesives, and surface coatings. The terms polymer and resin are used synonymously in the chemical industry, but the terms plastics, elastomers, and fibers have specific meanings (Sec. 4.5), and it is incorrect to refer to all synthetic polymers as plastics. A plastic is a material that is formed or fabricated from a polymer, usually by causing it to flow under pressure. Thus if a polymer is molded, extruded, cast, machined, or foamed to a particular shape, which may include both supported and unsupported film, the polymer can be described as a plastic. Usually a plastic contains pigments and additives such as antioxidants, plasticizers, and stabilizers.

This chapter includes some of the chemistry of individual polymer manufacture but is intended more as a broad description of how to synthesize a polymer, how to influence its properties, and how these properties relate to end uses that affect our daily lives.

In 1977 the US polymer industry produced about 52 billion lb of polymers (Table. 4.1) compared with the figure of 250 billion lb (Sec. 1.5) for the output of the US organic chemical and polymer industry. The

Table 4.1 US Polymer Production, 1977 (Estimated)

Polymer Use	Billion lb.
Plastics	31.0
Fibers	9.0
Elastomers	5.5
Coatings	4.0
Adhesives	2.5
	52.0

latter figure refers to chemicals actually isolated before being subjected to another reaction. Thus we might conclude that the polymer industry consumed only about one-fifth of the chemical industry output. Such a conclusion ignores an important element of double counting implicit in the statistics. If 1 billion lb of ethylene and 3 billion lb of benzene are produced from naphtha and are reacted together to give about 4 billion lb of ethylbenzene that is then dehydrogenated to a like amount of styrene that in turn is polymerized to almost 4 billion lb of polystyrene, then the production statistics will record 16 billion lb of chemicals. It is difficult to eliminate this element of double counting altogether, but one estimate is that the 52 billion lb of polymers consume about 128 billion lb of chemicals—about 51% of the total of all chemicals produced. To that should be added such materials as solvents for surface coatings, plasticizers for poly(vinyl chloride), as well as many compounding and processing aids. Thus we can say with confidence that the polymer industry consumes over half the tonnage output of the organic chemical industry.

Polymers may be subdivided into two categories, thermoplastic and thermosetting. Thermoplastics soften or melt when heated and will dissolve in suitable solvents. They consist of long chain molecules often without any branching (e.g., high density polyethylene). Even if there is branching (e.g., low density polyethylene) the polymer may still be two dimensional. Thermoplastics may be used in the five main applications of polymers—plastics, fibers, elastomers, coatings, and adhesives—as shown in Table 4.2. These are discussed further in Section 4.6.

Thermosets decompose on heating and are infusible and insoluble. They have elaborately cross-linked three-dimensional structures and are used for plastics, elastomers (lightly cross-linked), coatings, and adhesives but not fibres, which require unbranched linear molecules that can be suitably oriented (Part II, Sec. 3.2) during the spinning and drawing processes.

Table 4.2   Major Applications of Polymers

Plastics

Extruded products

Low density polyethylene
Poly(vinyl chloride)
Polystyrene and styrene copolymers
High density polyethylene
Poly(ethylene terephthalate)
Polypropylene
Acrylonitrile-butadiene-styrene
  copolymers
Cellulose acetate
Cellulose acetate butyrate

Molded products

Polystyrene and styrene copolymers
High density polyethylene
Polypropylene
Low density polyethylene
Poly(vinyl chloride)
Phenolics
Urea-formaldehyde
Melamine-formaldehyde
Acrylics
Cellulose acetate
Cellulose acetate butyrate

Film and Sheet

Low density polyethylene
Poly(vinyl chloride)
Regenerated cellulose
Acrylics
Poly(ethylene terephthalate)
Polypropylene
High density polyethylene

Foams

Polyurethane
Polystyrene

Fibers

Poly(ethylene terephthalate)
Nylon
Polyacrylonitrile copolymers
Polypropylene
Rayon
Cellulose acetate
Glass

Elastomers

Styrene-butadiene rubber
Polyisoprene
Ethylene-propylene terpolymers
Polybutadiene
Butadiene-acrylonitrile copolymers
Silicone
Sulfochlorinated polyethylene

Coatings

Paper and Textile applications

Low density polyethylene
Polystyrene and styrene copolymers
Poly(vinyl chloride)
Poly(vinyl acetate)
Urea-formaldehyde
Melamine-formaldehyde

Conventional coatings

Alkyds
Oils
Acrylics
Poly(vinyl acetate)
Poly(vinyl chloride)
Rosin, esters, adducts
Epoxy
Cellulose acetate
Cellulose acetate butyrate
Urea-formaldehyde
Melamine-Formaldehyde
Urethane
Polystyrene and styrene copolymers
Unsaturated polyesters

Adhesives

Laminating

Phenol-formaldehyde
Urea-formaldehyde
Melamine-formaldehyde

Conventional

Phenol-formaldehyde
Urea-formaldehyde
Melamine-formaldehyde
Poly(vinyl acetate)
Epoxy

Table 4.3 Production of Polymers for Plastics Applications, 1977

Polymer	Thermoplastic (A) or Thermoset (B)	Production (Million lb) United States	Production (Million lb) United Kingdom
1. Polyethylene (low density)	A	6470	919
2. Poly(vinyl chloride) and other vinyl resins	A	5250	930
3. Polystyrene and copolymers	A	4630	352
4. Polyethylene (high density)	A	3650	264
5. Polypropylene	A	2750	407
6. Phenol-formaldehyde	B	1460	132
7. Unsaturated polyester	B	1060	128
8. Acrylonitrile-butadiene-styrene (ABS)	A	900	79
9. Urea-formaldehyde	B	960	299
10. Epoxy	B	260	28
11. Nylon	A	200	44
12. Melamine-formaldehyde	B	200	56
13. Styrene-acrylonitrile (SAN)	A	150	—
14. All other resins[a]	A	2000	870

[a] Includes acetal, polysulfone, cellulosics, coumarone-indene, acrylics, fluorocarbons, alkyds, thermoplastic polyesters, polycarbonate, and modified poly(phenylene oxide).

Table 4.3 shows the sales of polymers for plastics applications in the United States and United Kingdom in 1977. In the United States the largest volume plastic is low density polyethylene whereas in the United Kingdom it is poly(vinyl chloride) (PVC). If we add together low and high density polyethylene, then polyethylene emerges as the most important plastic worldwide.

Sales of thermoplastics for plastics applications are about four times those of thermosets in spite of the fact that thermosets, especially urea-formaldehyde and phenol-formaldehyde resins, have been produced commercially for much longer than any of the thermoplastics. The thermosets have been unable to share more extensively in the phenomenal growth of plastics because they are difficult to process and do not lend themselves to the high production speeds that can be achieved, for example, by modern injection molding machines. Method of processing plastics are summarized in the notes to Part II, Chapter 2.

4.1 POLYMERIZATION

Before considering polymer properties let us describe how molecules link together to form polymers. There are two types of polymerization: addition or chain growth (also called simply chain) polymerization and condensation or step growth (also called simply step) polymerization. The terms chain and step are more accurate than the older terms, addition and condensation. Chain growth polymerization often involves monomers containing a carbon-carbon double bond, although cyclic ethers such as ethylene and propylene oxides and aldehydes such as formaldehyde polymerize in this way. Chain growth polymerization is characterized by the fact that the intermediates in the process—free radicals, ions, or metal complexes—are transient and cannot be isolated.

Step growth polymerization occurs because of reactions between molecules containing functional groups, for example, the reaction between a glycol and a dibasic acid to give a polyester.

$$HOCH_2CH_2OH + HOOC-\langle\bigcirc\rangle-COOH$$

Ethylene glycol   Terephthalic acid

$$\downarrow$$

$$HOCH_2CH_2O\,OC-\langle\bigcirc\rangle-COOH + H_2O$$

$$\downarrow \quad HOCH_2CH_2OH$$

$$HOCH_2CH_2O\,OC-\langle\bigcirc\rangle-CO\,OCH_2CH_2OH + H_2O$$

$$\downarrow \quad HOOC-\langle\bigcirc\rangle-COOH$$

$$HOCH_2CH_2O\,OC-\langle\bigcirc\rangle-CO\,OCH_2CH_2O\,OC-\langle\bigcirc\rangle-COOH + H_2O$$

$$\downarrow \text{ Continued condensation}$$

$$HOCH_2CH_2O-\left[OC-\langle\bigcirc\rangle-CO\,OCH_2CH_2O\right]_n-OC-\langle\bigcirc\rangle-COOH$$

Poly(ethylene terephthalate)

The low molecular weight intermediates are called oligomers, a term also used for the low molecular weight products obtained by chain growth polymerization. In the polyesterification shown here an oligomer can have two terminal hydroxyl groups, two terminal carboxyls, or one of each. The hydroxyls can react further with terephthalic acid and the carboxyls further with ethylene glycol. Alternatively, two oligomers can condense. The continuation of these reactions familiar from simple esterification chemistry, leads to the final polymer. The step growth or condensation reactions can be stopped at any time and low molecular weight polyesters (terminated by hydroxyl or carboxyl groups) isolated. Step growth polymerization, as opposed to chain growth polymerization, is therefore defined as a polymerization in which the intermediates can be isolated.

Usually a small molecule such as water is given off, but this is not always so. In the polymerization of the cyclic monomer caprolactam, for example, both functional groups are in the same molecule, and there is no by-product. Indeed, a mole of water is needed to start the polymerization. Because there is no small molecule given off the reaction reaches an equilibrium in which about 10% of the caprolactam remains unreacted. The monomer and oligomers (see below) that are always present must be removed by washing with water. This polymerization is carried out under the same conditions as are used to produce nylon from two bifunctional reagents and is clearly a step growth reaction. It should be noted, however, that it is the polymerization of 6-aminocaproic acid, not of caprolactam itself.

$$n\ \overline{HNCH_2CH_2CH_2CH_2CH_2CO} \xrightarrow{\ H_2O\ } n\ H_2NCH_2CH_2CH_2CH_2CH_2COOH$$

Caprolactam   6-Aminocaproic acid

$$\longrightarrow\ H \left[NHCH_2CH_2CH_2CH_2CH_2CO \right]_n OH$$

Nylon 6

Caprolactam can also be polymerized by a chain mechanism using ionic initiators. Figure 4.1 demonstrates the ionic polymerization of caprolactam with sodium methoxide as the initiator. Ionic polymerization is discussed in greater detail in Section 4.3.6.

4.2 FUNCTIONALITY

Functionality is a measure of the number of linkages one monomer may form with another. A monomer that, when polymerized, may join with two other monomers is termed bifunctional. If it may join with three or more molecules it is tri- or polyfunctional. Glycols and dibasic acids are clearly bifunctional. Similarly 6-aminocaproic acid, the reaction product

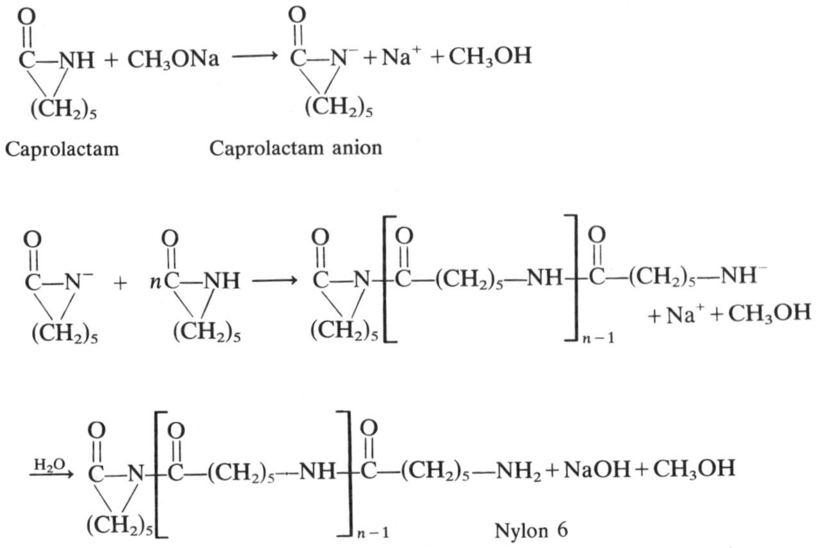

$$\underset{\substack{\diagdown \diagup \\ (CH_2)_5}}{\overset{\overset{\displaystyle O}{\|}}{C}-NH} + CH_3ONa \longrightarrow \underset{\substack{\diagdown \diagup \\ (CH_2)_5}}{\overset{\overset{\displaystyle O}{\|}}{C}-N^-} + Na^+ + CH_3OH$$

Caprolactam   Caprolactam anion

$$\underset{\substack{\diagdown \diagup \\ (CH_2)_5}}{\overset{\overset{\displaystyle O}{\|}}{C}-N^-} + n\underset{\substack{\diagdown \diagup \\ (CH_2)_5}}{\overset{\overset{\displaystyle O}{\|}}{C}-NH} \longrightarrow \underset{\substack{\diagdown \diagup \\ (CH_2)_5}}{\overset{\overset{\displaystyle O}{\|}}{C}-N}\!\!\left[\overset{\overset{\displaystyle O}{\|}}{C}-(CH_2)_5-NH\right]_{n-1}\!\!\overset{\overset{\displaystyle O}{\|}}{C}-(CH_2)_5-NH^- + Na^+ + CH_3OH$$

$$\xrightarrow{H_2O} \underset{\substack{\diagdown \diagup \\ (CH_2)_5}}{\overset{\overset{\displaystyle O}{\|}}{C}-N}\!\!\left[\overset{\overset{\displaystyle O}{\|}}{C}-(CH_2)_5-NH\right]_{n-1}\!\!\overset{\overset{\displaystyle O}{\|}}{C}-(CH_2)_5-NH_2 + NaOH + CH_3OH$$
Nylon 6

Figure 4.1   Ionic polymerization of caprolactam.

of water and caprolactam, is bifunctional because it contains a carboxyl and an amino group. The functionality rules state that if bifunctional molecules react, a linear polymer will be obtained. If one reactant has a functionality greater than 2 a cross-linked structure results (see notes).

The ethylenic double bond so important in chain polymerization has a functionality of 2 even though the organic chemist would regard the double bond as one functional group. However the ability of the extra electron pair of the ethylenic linkage to enter into the formation of two bonds makes it bifunctional. Thus ethylene polymerizes to form a linear thermoplastic polymer. The double bond in propylene contributes a functionality of 2, but propylene also possesses allylic hydrogens that are activated by peroxide initiators so that a cross-linked structure results. Thus propylene has a functionality greater than 2 toward peroxide catalysts. On the other hand, Ziegler-Natta catalysts do not activate the allylic hydrogens. Propylene shows a functionality of 2 toward them, and a linear polymer results. This chemistry is harnessed in some ethylene-propylene rubbers. A linear ethylene-propylene copolymer is made with a Ziegler-Natta catalyst. It may then be cross-linked with a peroxide catalyst (Part II, Sec. 4.2.3).

In the production of unsaturated polyester resins a linear liquid polymer is made by step growth polymerization typically of propylene glycol with phthalic and maleic anhydrides. Each of these reagents exhibits functionalities of 2 in an esterification reaction. The maleic anhydride however has a double bond that can undergo chain growth polymerization. Thus subsequent treatment of the unsaturated polyester with styrene and a peroxide catalyst leads to a solid, infusible thermoset in which polyester chains are cross-linked by polystyrene chains. The maleic anhydride has a functionality of 2 in both the step growth and chain growth polymerizations. In chain growth polymerization, cross-linking results because the unsaturated polyester with its multiple double bonds has a functionality much greater than 2.

The fact that monomers exhibit different functionalities toward different reagents and polymerization techniques provides a means by which an initial polymerization can give a linear polymer that can subsequently be cross-linked by a different technique. Further examples of this are given in Section 4.4.

Conjugated structures such as those in butadiene are considered to have a functionality of only 2. When they polymerize, linear polymers are formed that still contain double bonds. These can subsequently be

$$n\,H_2C=CH-CH=CH_2 \rightarrow \{CH_2-CH=CH-CH_2\}_n$$

Butadiene   Polybutadiene

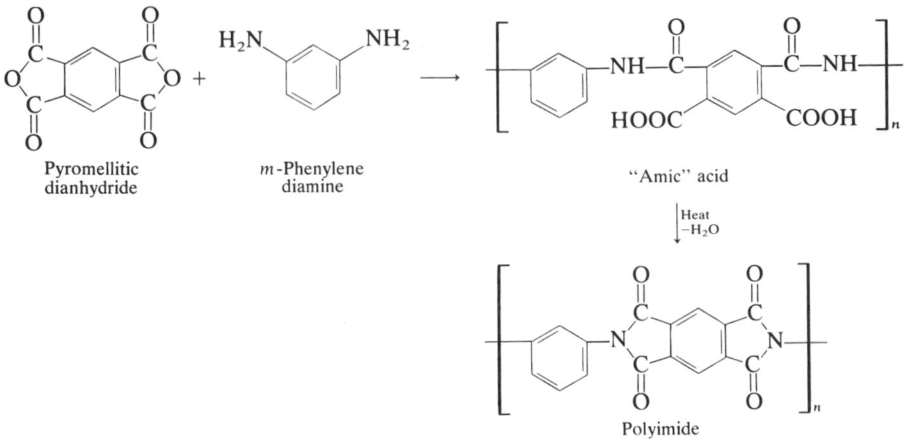

Figure 4.2 Polymide synthesis.

reacted to form a cross-linked polymer, but in the initial polymerization butadiene is bifunctional.

Pyromellitic dianhydride has a functionality of two when reacted with diamines, and linear polyimides result as shown in Figure 4.2.

On the other hand, this anhydride reacts with water to form a tetracarboxylic acid that has a functionality of 4 when reacted with compounds containing hydroxyl groups.

The functionality of a molecule is not always completely obvious. The situation with a double bond has already been discussed. Formaldehyde has a functionality of 2 and will polymerize to the common laboratory reagent, paraformaldehyde:

$$n \; H-\overset{\overset{\displaystyle H}{|}}{C}=O \; \longrightarrow \; -\!\!\left[CH_2O\right]\!\!-_n$$

Although very high molecular weights can be obtained, the product is of little commercial value because it "unzips" easily to regenerate formaldehyde. This can be prevented and practical polymers obtained by acetylation of the hydroxyl end groups. Alternatively a formaldehyde-ethylene glycol copolymer is made and is then subjected to conditions that would normally degrade it. The formaldehyde groups at the end of the chains "peel off" until an ethylene glycol unit is encountered. Depolymerization ceases, and a stable polymer molecule with hydroxyethyl end groups is left with the following structure:

$$HOCH_2-CH_2-O \, (CH_2O)_n \, CH_2-CH_2OH$$

Formaldehyde polymers are known as polyacetals. Aldehydes in general

a   —A—A—A—A—A—A—

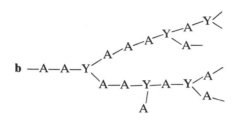

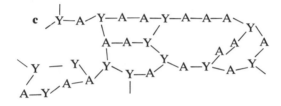

Figure 4.3   Linear, Branched and crosslinked structures.

can be polymerized anionically or cationically (Sec. 4.3.6) to give polymers with a ᴡᴡC—O—C—O—C—Oᴡᴡ backbone.

If monomers have functionalities greater than 2, chain branching and eventually cross-linking can occur. It can be seen intuitively that bifunctional monomers can only form linear chains (**a**, Fig. 4.3). If a trifunctional monomer (Y in **b**, Fig. 4.3) is added, chain branching can occur, and if there is sufficient of it an elaborate three-dimensional network can result. If some of the chains attached to the Y groups in **c** of Figure 4.3.2 are thought of as coming off at right angles to the plane of the paper some concept of the three-dimensional structure can be gained.

Glycerol has a functionality of 3 and if condensed with a dibasic acid can give the multifunctional oligomers shown. Cross-linked polymers soon become insoluble and infusible as their molecular weight builds up. The cross-linking may lead to excellent strength characteristics, but the infusibility and insolubility means that such polymers are difficult to fabricate. The chemistry of thermoset polymers produced by step growth polymerization is further discussed in Section 4.4.

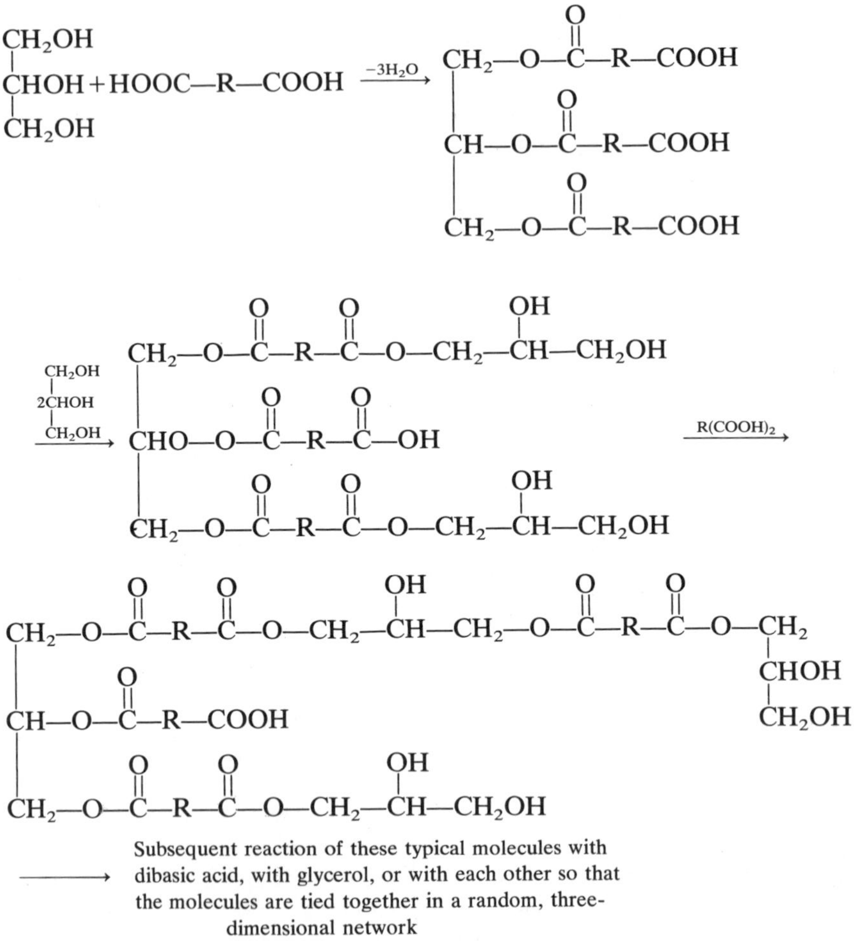

$$
\begin{array}{l}
\text{CH}_2\text{OH}\\
|\\
\text{CHOH} + \text{HOOC—R—COOH} \xrightarrow{-3\text{H}_2\text{O}}\\
|\\
\text{CH}_2\text{OH}
\end{array}
\quad
\begin{array}{l}
\overset{\displaystyle O}{\overset{\|}{}}\\
\text{CH}_2\text{—O—C—R—COOH}\\
\overset{\displaystyle O}{\overset{\|}{}}\\
\text{CH—O—C—R—COOH}\\
\overset{\displaystyle O}{\overset{\|}{}}\\
\text{CH}_2\text{—O—C—R—COOH}
\end{array}
$$

$$
\begin{array}{l}
\text{CH}_2\text{OH}\\
2\text{CHOH}\\
\text{CH}_2\text{OH}
\end{array}
\longrightarrow
\begin{array}{l}
\text{CH}_2\text{—O—C—R—C—O—CH}_2\text{—CH—CH}_2\text{OH}\\[4pt]
\text{CHO—O—C—R—C—OH}\\[4pt]
\text{CH}_2\text{—O—C—R—C—O—CH}_2\text{—CH—CH}_2\text{OH}
\end{array}
\xrightarrow{\text{R(COOH)}_2}
$$

(with O double bonds on each C and OH on the CH of the CH₂OH chains)

$$
\begin{array}{l}
\text{CH}_2\text{—O—C—R—C—O—CH}_2\text{—CH—CH}_2\text{—O—C—R—C—O—CH}_2\\
\hspace{16em}\text{CHOH}\\
\text{CH—O—C—R—COOH}\hspace{10em}\text{CH}_2\text{OH}\\
\text{CH}_2\text{—O—C—R—C—O—CH}_2\text{—CH—CH}_2\text{OH}
\end{array}
$$

Subsequent reaction of these typical molecules with dibasic acid, with glycerol, or with each other so that the molecules are tied together in a random, three-dimensional network

4.3 STEP GROWTH AND CHAIN GROWTH POLYMERIZATIONS

Step growth polymerization can be described as a simple chemical reaction carried out repeatedly. Polyesterification, for example, is brought about by the same catalysts as esterification reactions, and the equilibrium is pushed to the ester side of the equation by removal of the by-product water either by simple distillation or as an azeotrope.

A major difference between a simple condensation reaction and a polycondensation is that the high molecular weight of the polymer product increases the viscosity of the reaction mixture if the polymer is soluble in it; if not, it precipitates. To solve the viscosity problem, the reaction may be carried out in a solvent, a technique that is particularly useful if the polymer is to be used in a surface coating that requires solvent. More often the engineer is called on to devise equipment with powerful stirrers that can accommodate viscous masses. In the production of many thermoset polymers, polymerization is interrupted at an early stage before cross-linking starts. The product is still fusible and soluble and is known as a "B-stage" polymer. *In situ* curing, usually with the aid of a catalyst, is relied upon to build up molecular weight and achieve the cross-linked state. Phenolics are often used as B-stage polymers (Sec. 4.4). In polyimide formation an intermediate chemical species, an "amic" acid, is formed. This is soluble, albeit in very strong solvents such as dimethylformamide. The solution can, however, be laid down as film and then heated further to achieve polyimide formation.

The molecular weight and tendency to gel of a polycondensation polymer may be controlled by addition of a monofunctional compound known as a "chain stopper." In the production of polymeric plasticizers such as poly(ethylene glycol adipate) (Part II; Sec. 11.1), for example, butanol is used as a chain stopper. Chain stoppers are also important in the production of alkyds (Part II, Sec. 5.6).

Chain growth polymerization proceeds rapidly by way of transient intermediates to give the final polymer. We can write an overall equation,

$$n \; CH_2{=}CHX \longrightarrow {+}CH_2{-}CHX{+}_n$$

but it provides no indication of the reaction mechanism. Polymerization is started by a chain initiator that converts a molecule of monomer into a free radical or an ion or else by a catalyst that converts the monomer to a metal complex. The free radicals or ions then undergo so-called propagation reactions that build up the polymer chain. In the case of metal complex catalysis, often referred to as Ziegler-Natta catalysis, the propagation takes place on the surface of the metal complex catalyst. Finally, there must be a chain termination step in which the transient intermediate, now a polymer chain, is stabilized.

In chain growth polymerization repeating units are added one at a time, as opposed to step growth polymerization where oligomers may condense with one another. Propagation and termination steps are very rapid. Once a chain is initiated monomer units add on to the growing chain very quickly, and the molecular weight of that unit builds up in a fraction of a second. Consequently the monomer concentration decreases steadily throughout the reaction. Prolonged reaction time has little effect on molecular weight but does provide higher yields. At any given time the reaction mixture contains unchanged reactant and "fully grown" polymer chains but a very low concentration of growing chains. The growing chains cannot readily be separated from the reaction mixture.

In step growth polymerization (Sec. 4.1) the monomer does not decrease steadily in concentration; rather it disappears early in the reaction because of the ready formation of low molecular weight oligomers. The molecular weight of a given polymer chain increases continually throughout the reaction, and thus long reaction times build up the molecular weight. After the early stages of the reaction there is neither much reactant nor a great deal of "fully grown" polymer present. Instead there is a wide distribution of slowly growing oligomers. If desired this distribution can be calculated and the separate oligomers isolated from the reaction mixture.

4.3.1 FREE RADICAL POLYMERIZATION

Free radical polymerization is initiated by free radicals from compounds such as benzoyl peroxide which, on heating, decomposes to give benzoyl-peroxy radicals, some of which eliminate carbon dioxide to give phenyl radicals. One of the free radicals then adds on to a molecule of monomer such as ethylene, vinyl chloride, or styrene to convert that monomer to a radical. Initiation is now complete, and the initiating free radical is incorporated into one end of a polymer chain. Its concentration in a high molecular weight polymer is so small that it does not affect final properties. The radical now reacts with another molecule of monomer to give a larger free radical, and this chain propagation process continues until the chain is terminated.

Initiator formation: $I \longrightarrow 2R\cdot$

Initiation: $R\cdot + CH_2{=}CH \longrightarrow RCH_2{-}CH\cdot$
$\qquad\qquad\qquad\quad | \qquad\qquad\qquad\quad |$
$\qquad\qquad\qquad\quad X \qquad\qquad\qquad\quad X$

Propagation: $RCH_2CH\cdot + CH_2{=}CH \longrightarrow RCH_2CHCH_2CH\cdot$
$\qquad\qquad\qquad | \qquad\qquad\quad | \qquad\qquad\quad | \qquad\quad |$
$\qquad\qquad\qquad X \qquad\qquad\quad X \qquad\qquad\quad X \qquad\quad X$

Continued propagation: 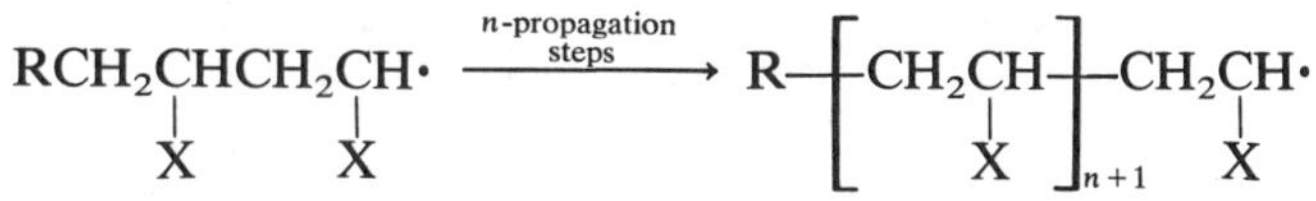

Termination: Radical coupling

Disproportionation

Chain transfer (termination or branching—Sec. 4.3.2)

What can stop the chain? The three possible processes are called coupling, disproportionation, and chain transfer. Coupling occurs when two growing free radicals collide head to head to form a single stable molecule with a molecular weight equal to the sum of the individual molecular weights. In the disportionation reaction two radicals again meet, but this time a proton transfers from one to the other to give two stable molecules, one saturated and the other with a terminal double bond. Above 60°C polystyrene terminates predominantly by coupling whereas poly(methyl methacrylate) terminates entirely by disproportionation. At lower temperatures both processes occur.

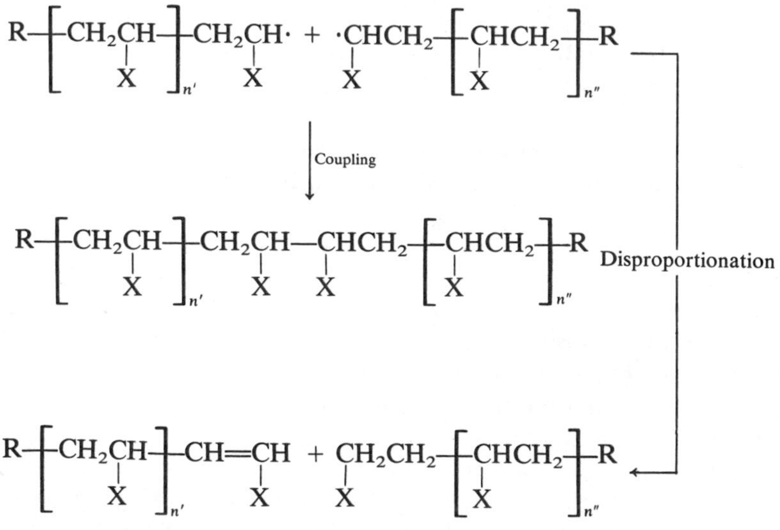

Chain transfer can cause either termination or branching. It is discussed in Section 4.3.2.

Propagation reactions in chain growth polymerizations are very fast, and chains build up in less than a second. Indeed polymerizations can and sometimes do become explosive. Termination steps occur rarely relative to the propagation reaction, not because they are slow but because the concentrations of free radicals are normally so low that encounters between them are rare.

In the polymerization of ethylene it makes no difference which end of the molecule is attacked by the free radical. With unsymmetrical monomers such as vinyl chloride or styrene, however, there could be head-to-tail propagation (eq. **a**), head-to-head propagation, or completely random addition. In the head-to-head case the side chain or hetero-atom X would sometimes occur on adjacent carbon atoms (eq. **b**).

$$\textbf{a} \quad \sim\sim\sim CH_2-\underset{\underset{X}{|}}{C}H\cdot + CH_2=\underset{\underset{X}{|}}{C}H \longrightarrow \sim\sim\sim CH_2-\underset{\underset{X}{|}}{C}H-CH_2-\underset{\underset{X}{|}}{C}H\cdot \longrightarrow etc.$$

$$\textbf{b} \quad \sim\sim\sim CH_2-\underset{\underset{X}{|}}{C}H\cdot + \underset{\underset{X}{|}}{C}H=CH_2 \longrightarrow R-CH_2-\underset{\underset{X}{|}}{C}H-\underset{\underset{X}{|}}{C}H-CH_2\cdot \longrightarrow etc.$$

Head-to-head propagation rarely occurs because the unpaired electron in the free radical prefers to locate itself on the —CHX end of the monomer molecule where it has a better opportunity to delocalize. The free radical is thus more stable. Head-to-tail polymerization is the norm, and only an occasional monomer molecule slips in the wrong way. Termination by coupling, of course, creates a head-to-head structure.

The relative rates of the initiation, propagation, and termination processes are reflected in the key property of molecular weight on which many of the other properties of the polymer depend. If the rate of initiation is high, for example, then the concentration of free radicals at any given moment will be high, and they will stand a good chance of colliding and coupling or disproportionating. A high initiation rate will therefore lead to a low molecular weight polymer.

For a high molecular weight polymer, a low initiation rate is required together with a high propagation rate. We might also say a low termination rate, but because termination steps have no activation energy they occur on every collision and are diffusion-controlled. The termination rate is decreased by increase in viscosity or decrease of concentration in a system. If propagation and termination steps have comparable rates a polymer will not result. A propagation rate thousands of times the termination rate is required, and the molecular weight is in fact a function of the ratio of propagation to termination rates.

4.3.2 CHAIN TRANSFER

Another factor that affects molecular weight is chain transfer. A growing polymer radical may extract a hydrogen atom from a finished polymer chain. This "finished" polymer chain now becomes a radical and starts to grow again. If the hydrogen atom is extracted from the end of the chain the new chain simply continues to grow linearly. But if, as is more probable statistically, the hydrogen atom is extracted from the body of the chain, then further propagation occurs at right angles to the original polymer chain and a branch forms.

$$\sim\sim CH_2-CH\sim\sim + \sim\sim CH^{\bullet} \longrightarrow \sim\sim\sim CH_2-\overset{\bullet}{C}\sim\sim + \sim\sim CH_2$$

| X | X | X | X |

"Finished" polymer chain   Growing chain

$$\Big| CH_2CHX$$

$$\sim\sim CH_2-\overset{(CH_2CHX)_n CH_2CHX^{\bullet}}{\underset{\diagdown}{\overset{\diagup}{CX}}} \xleftarrow{nCH_2CHX} \sim \sim CH_2-CX\overset{\diagup}{\underset{\diagdown}{\quad}}\overset{CH_2-CH^{\bullet}}{\underset{X}{\diagup}}$$

Branched polymer

Branching can have a marked effect on polymer properties. For example, it can keep polymer molecules from achieving nearness and hence reduce cohesive forces between them. Branching can also make it harder for polymer crystals to form (Sec. 4.5.1). The following equation shows how branching takes place in polyethylene.

$$\sim\sim\overset{\overset{\displaystyle CH_2}{\diagup\;\diagdown}}{\underset{\underset{\displaystyle H}{|}}{CH}}\quad\overset{\overset{\displaystyle CH_2}{|}}{\underset{\underset{\displaystyle \cdot CH_2}{|}}{CH_2}} \longrightarrow \sim\sim\overset{\overset{\displaystyle CH_2}{\diagup\;\diagdown}}{\underset{\cdot}{CH}}\quad\overset{\overset{\displaystyle CH_2}{|}}{\underset{\underset{\displaystyle CH_3}{|}}{CH_2}} \xrightarrow{C_2H_4} \sim\sim\overset{\overset{\displaystyle C_4H_9}{|}}{CHCH_2CH_2\cdot}$$

Pseudo 6-membered ring

etc.

$$\sim\sim\overset{\overset{\displaystyle C_4H_9}{|}}{CHCH_2CH_2}\sim\sim$$

A growing polymer molecule with a free radical end can bend to form a pseudo 6-membered ring that facilitates the transfer of the free radical site from the end of the chain to a carbon atom within the chain. The chain then starts to grow from this new site with the net result that the branch has four carbon atoms. Low density polyethylene is indeed characterized by C_4 branches.

Chain transfer can occur not only to another polymer chain but also to a molecule of monomer. The new radical will then propagate in the usual way. If this happens often, a low molecular weight polymer will be formed.

Chain transfer is undesirable except when it is used intentionally to limit molecular weight. It can be controlled by addition of chain transfer agents. These are materials from which hydrogen atoms can readily be abstracted. If a growing radical is liable to extract a hydrogen atom it will do so preferentially from the chain transfer agent rather than another polymer molecule. The problem of branching will thus be avoided but not that of reduced molecular weight. Dodecyl mercaptan is used as a chain transfer agent in low density polyethylene and rubber polymerizations. When it loses a hydrogen atom a stable disulfide forms. The formation of a stable compound after loss of a proton is a key characteristic of a chain transfer agent.

$$C_{12}H_{25}SH \longrightarrow 2H + C_{12}H_{25}S\text{---}SC_{12}H_{25}$$

Dodecyl mercaptan   Dodecyl disulfide

Phenols may be used similarly because they give up their phenolic hydrogen readily and the resulting phenoxy radical is relatively stable and does not add to monomer.

Hydroquinone   Phenoxy radical

During storage, monomers are sometimes stabilized with polyhydric phenols (e.g., hydroquinone and *t*-butylcatechol) or aromatic amines (e.g., methylene blue) so that they do not polymerize spontaneously.

Methylene blue

Hydroquinone is ineffective in the absence of oxygen, so its mode of action is probably more complicated than suggested here. If a free radical should appear in the monomer it immediately accepts a proton from the inhibitor and is "squelched." When it is desired to polymerize the monomer the inhibitor must be removed, usually by distillation.

4.3.3 COPOLYMERIZATION

Discussion of propagation rates leads to the topic of copolymerization. Copolymers are polymers made from two or more monomers and are of two kinds. In regular copolymers the monomer units are arranged alternately. Many step growth copolymers are regular. An example of a regular chain growth copolymer is one based on maleic anhydride and styrene. Maleic anhydride will not homopolymerize but reacts rapidly in a polymerization reaction with another monomer. Thus one styrene monomer will react with a maleic anhydride monomer. The resulting free radical,

*R = Initiator residue

has a choice of reacting further with another styrene molecule or with another maleic anhydride. It chooses the styrene because the maleic anhydride will not react with itself. The free radical,

chooses a maleic anhydride monomer because the reaction rate between styrene and maleic anhydride is much greater than between styrene-styrene. In this way a completely regular copolymer, $\sim$ S—MA—S—MA—S$\sim$, is obtained.

In the other type of copolymer, the monomer units are not in an orderly sequence but are random. To form a random copolymer, the two monomers must react with themselves at a rate comparable to that at which they react with each other. If the propagation rates differ widely, the first polymer molecules to be formed will consist almost entirely of the fast reacting monomer, and when all of it is used up the slow reacting material will polymerize to give a polymer consisting almost entirely of the slow reacting material. The possible propagation reactions are as follows. It is the relative rates of these processes that decide whether a random copolymer, two homopolymers, or something in between is obtained.

$$\textbf{a} \quad \text{\wavy}A\cdot + A \longrightarrow \text{\wavy}A-A\cdot$$

$$\textbf{b} \quad \text{\wavy}A\cdot + B \longrightarrow \text{\wavy}A-B\cdot$$

$$\textbf{c} \quad \text{\wavy}B\cdot + B \longrightarrow \text{\wavy}B-B\cdot$$

$$\textbf{d} \quad \text{\wavy}B\cdot + A \longrightarrow \text{\wavy}B-A\cdot$$

Copolymerization serves several functions. First, a copolymerizing monomer may be included to plasticize the polymer, that is to make it softer. Because vinyl acetate gives too brittle a film for water-borne paints (Part II, Sec. 5.5.2), it may be polymerized with 2-ethylhexyl acrylate. Second, the copolymerizing monomer may insert functional groups. In unsaturated polyesters (Sec. 4.2) the maleic anhydride provides double bonds that may subsequently be cross-linked by chain growth polymerization. In elastomers, a comonomer with two double bonds is almost always used. One double bond engages in chain growth polymerization, and the other remains intact on each recurring unit so that sites for "vulcanization" or cross-linking are present. Thus butyl rubber (Part II, Sec. 4.2.2) is a copolymer of isobutene with a small amount of isoprene.

Finally, copolymerization can be used to reduce crystallinity (Sec. 4.5.1). Low density polyethylene is about 50% crystalline. By making a copolymer with propylene this crystallinity is destroyed, and a polymer results that becomes an elastomer on cross-linking.

There are, in all, four types of copolymers. In regular copolymers the monomer units alternate in the chain (—A—B—A—B—A—B—A—B—); in random copolymers they follow each other indiscriminately (—A—A—B—A—B—B—A—B—). Some copolymers consist of a group of one polymerized monomer followed by a group of the other (—A—A—A—A—B—B—B—B—), and these are called block copolymers (Sec. 4.3.8). A fourth copolymer is the graft copolymer, which results when a polymer chain of one monomer is grafted on to an existing polymer backbone (Sec. 4.3.9):

$$
\begin{array}{c}
\text{B—B—B—B—} \\
| \\
\text{—A—A—A—A—A—A—}
\end{array}
$$

4.3.4 MOLECULAR WEIGHT

We have referred several times to the molecular weight of a polymer. This is not as simple a concept as it sounds. Since the chains in a sample of polymer do not all have the same number of recurring units, the molecular weight of a polymer is always an average. Actually a broad molecular weight distribution is often desirable, for oligomers may serve as lubricants during processing and as plasticizers thereafter.

Molecular weight of polymers is commonly expressed in two ways: by number average $\bar{M}_n$ and by weight average $\bar{M}_w$. The number average is obtained by adding the molecular weights of all the molecules and dividing by the number of molecules. If we have n_1 molecules of molecular weight M_1, n_2 of molecular weight M_2, and n_x of molecular weight M_x then

$$\bar{M}_n = \frac{n_1 M_1 + n_2 M_2 + \cdots n_x M_x + \cdots}{n_1 + n_2 + \cdots n_x \cdots}$$

The weight average, on the other hand, is calculated according to the weight of all the molecules at each molecular weight. Let w_1 be the weight (in molecular weight units) of molecules of molecular weight M_1, w_2 the weight of molecules of molecular weight M_2 and so on, then

$$\bar{M}_w = \frac{w_1 M_1 + w_2 M_2 + \cdots w_x M_x + \cdots}{w_1 + w_2 + w_3}$$

But the total weight of all molecules with molecular weight w_1 is $n_1 M_1$, so we can substitute $w_1 = n_1 M_1$, $w_2 = n_2 M_2$, $w_x = n_x M_x$, and so on in the above equation, whence

$$\bar{M}_w = \frac{n_1 M_1^2 + n_2 M_2^2 + \cdots n_x M_x^2 \cdots}{n_1 M_1 + n_2 M_2 + \cdots n_x M_x \cdots}$$

$\bar{M}_n$ tells us where most of the polymer molecules are relative to the molecular weight distribution. $\bar{M}_w$, on the other hand, tells us where most of the weight is regardless of the molecular weight distribution. Because $\bar{M}_w$ is biased towards molecules with higher molecular weight, it will be larger than $\bar{M}_n$.

As an example consider three persons, two weighing 100 lb and one weighing 200 lb. Their number average weight is $(100 + 100 + 200)/3 = 133\frac{1}{3}$ lb, but their weight average is $(100^2 + 100^2 + 200^2)/(100 + 100 + 200) = 150$ lb. In the first instance we can consider that a person was selected at random; in the second, that a pound of weight was selected at random. The second selection (weight average) will naturally lead to a higher result because the pound of weight will tend to be selected from the heavier persons.

$\bar{M}_n$ and $\bar{M}_w$ both provide a narrow view of molecular weight, but their ratio, $\bar{M}_w/\bar{M}_n$, will tell us something about the molecular weight distribution. If $\bar{M}_w/\bar{M}_n = 1$ then all the molecules have the same molecular weight, and as the distribution of molecular weights becomes wider this ratio increases.

Boiling point elevation, freezing point depression, osmotic pressure, and end group analysis give number average molecular weight; light scattering and sedimentation methods give weight averages. Viscosity measurements give a value somewhere between the two.

The molecular weight profile of a polymer can be determined only by fractionation. Cumbersome solvent precipitation techniques give numerous fractions, and the molecular weight of each is determined. The fractions must be so narrow that for each of them $\bar{M}_w/\bar{M}_n$ is effectively unity.

4.3.5 POLYMERIZATION PROCEDURES

Chain growth polymerizations, whether initiated by free radicals as we have already described, or by ions or metal complexes as we describe later, are carried out by four different procedures—bulk, solution, suspension, and emulsion polymerization.

In bulk polymerization the monomer and the initiator are combined in a vessel and heated to the proper temperature. This procedure, although the simplest, is not always the best. The polymer that forms may dissolve in the monomer to give a viscous mass, and heat transfer becomes difficult. Heat cannot escape, and the polymer may char or develop voids. If the exotherm gets out of hand the system may explode.

Even so the polymerization of ethylene by the high pressure method is a bulk polymerization and is one of the polymerizations carried out on the largest scale. Fortunately the polymer does not dissolve in the monomer. Instead it collects in the bottom of the reactor and is drawn off. The exotherm still presents a problem, and the strictest possible control of temperature and heat transfer is necessary to prevent problems. The polymerization of methyl methacrylate to "Lucite" ("Plexiglas," "Perspex") is also carried out in bulk.

The other polymerization procedures are all designed to solve the problem of heat transfer. In solution polymerization the reaction is carried out in a solvent that acts as a heat sink and also reduces the viscosity of the reaction mixture. The snags with solution polymerization are first, it is frequently difficult to remove the last traces of solvent from the polymer, and second, the solvent may participate in chain transfer reactions so that low molecular weight polymers result. Solution polymerization is nonetheless used for the important polymerizations of ethylene to high density (low pressure) polyethylene in one version of the Phillips process (Sec. 4.3.11) and propylene to polypropylene. The monomer is dissolved in a solvent, and as the polymer forms, a polymer-catalyst slurry results. Initially it was necessary to separate the catalyst by a cumbersome process. This has led to the development of catalysts that separate more readily and to vapor phase processes that are basically bulk polymerizations. Also, catalysts are now available that give such high yields, and thus are present in such low concentrations, that they can be left in the polymer without affecting its properties.

In suspension polymerization the monomer and catalyst are suspended as droplets in a continuous phase such as water. These droplets have a high surface-to-volume ratio so heat transfer to the water is rapid. The droplets are maintained in suspension by continuous agitation and also, if necessary, by addition of a water-soluble polymer such as methylcellulose that increases viscosity of the water. Finely divided inorganic materials such as clay, talc, aluminium oxide, and magnesium carbonate have a similar stabilizing effect on the suspension. The need to remove these materials is one of the disadvantages associated with their use. Poly(vinyl chloride) is frequently made by suspension polymerization.

The final procedure is emulsion polymerization, a technique that was developed as part of the synthetic rubber program during World War II. As its name implies, it uses an emulsifying agent, usually various kinds of soap. In solution these form micelles (Part II, Sec. 7.1) in which the nonpolar hydrophobic ends of the soap molecules point inward, and the polar hydrophilic groups point outward and interact with the water. If monomer is added it is absorbed into the micelle to give a stable emulsion particle. If more monomer is added than can be absorbed in the micelles, a separate monomer droplet phase may form that is also stabilized by the soap molecules, the droplets being a micrometer or more in diameter. A water-soluble composite initiator called a "redox" catalyst is then added. This consists of a mixture of a reducing agent and an oxidizing agent. An example is ferrous ammonium sulfate and hydrogen peroxide. In the absence of monomer the former would reduce the latter in a two-stage process:

$$Fe^{2+} + H_2O_2 \longrightarrow Fe^{3+} + OH\cdot + OH^-$$
$$Fe^{2+} + OH\cdot \longrightarrow Fe^{3+} + OH^-$$

If monomer is present, however, the hydroxyl free radical can initiate polymerization. Other redox systems include benzoyl peroxide-ferrous ammonium sulfate, hydrogen peroxide-dodecyl mercaptan, and potassium persulfate-potassium thiosulfate which gives radical ions:

$$S_2O_8^= + S_2O_3^= \longrightarrow SO_4^= + SO_4^{\cdot} + S_2O_3^{\cdot}$$

Persulfate Thiosulfate
ion   ion

These polymerizations must be carried out with rigorous exclusion of oxygen, which is an inhibitor.

The free radicals diffuse into the micelles, and polymerization takes place within them. Diffusion into the droplets also occurs, but since they have a far lower surface-to-volume ratio than the micelles virtually none of the polymerization takes place within them. As polymer is formed, the micelles grow by diffusion of monomer from the droplets into the micelles. Rather than providing a site for polymerization the droplets serve as reservoirs for monomer that will later react in the micelles.

Polymerization within a micelle may take as long as 10 seconds. Very high molecular weights are produced, higher than by any of the three other procedures. The product is a latex, a dispersion of solid particles in water, which is frequently a desirable form for a polymer. For example, polyvinyl acetate or polyacrylate latices are used as such for "emulsion" paints. On the other hand if solid polymer is required the dispersion must be broken and the polymer precipitated.

There are two important differences between emulsion and suspension polymerization. In emulsion polymerization the catalyst or initiator is in the aqueous phase, not dissolved in the monomer, and the polymer particles produced are at least an order of magnitude smaller than those obtained from suspension polymerization.

4.3.6 IONIC POLYMERIZATION

Free radical initiation is the most widely used way to produce polymers (Fig. 4.4). A second method involves initiation by ions, either by anions or cations. Table 4.4 provides a list of ionic initiators useful for polymerization. In Table 4.5 there is a list of monomers and an indication as to whether they can be polymerized ionically or cationically. Many of them can be polymerized also by free radicals and by the metal complex catalysts discussed later. Ethylene may be polymerized cationically and also with the aid of free radicals. Propylene, however, has allylic hydrogens on the methyl group, and any attempt at free radical polymerization leads to low molecular weight cross-linked structures.

As a general rule monomers containing electron-withdrawing groups are more easily polymerized anionically, whereas those with electron-donating groups are more easily polymerized cationically. Nonetheless styrene, which contains the electron-withdrawing phenyl group, may be polymerized both anionically and cationically and, for that matter, by free radicals. Cationic polymerization of styrene, however, yields low molecular weight polymers.

Ionic polymerization is usually unsuitable for the preparation of

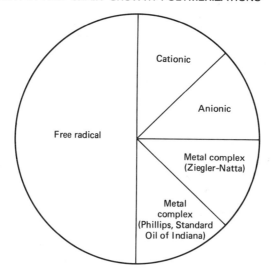

Figure 4.4 Methods of polymerization—use distribution.

copolymers. This is because the differences in the stabilities of organic ions are much greater than those between the corresponding radicals. It represents a serious limitation to ionic polymerization. An exception is block copolymers (Sec. 4.3.8), which may be prepared by ionic polymerization because the monomers are added successively not simultaneously.

The initiation step in anionic polymerization is the production of an anion from the monomer by a strong base. This is shown in the equation that follows where butyl lithium is the initiator. Butyl lithium and other anionic initiators such as sodium or potassium amides in liquid ammonia or sodium cyanide in dimethylformamide are expensive and not recoverable. Consequently this procedure is used only where there is no cheaper method of polymerization available and when the value of the product justifies the high initiator cost.

$$CH_2{=}\underset{X}{\overset{|}{C}}H + C_4H_9Li \longrightarrow C_4H_9CH_2\underset{X}{\overset{|}{C}}H^-Li^+$$

The propagation step in anionic polymerization is formally similar to that in free radical polymerization (Sec. 4.3.1), but actually there are differences. Ions like to be solvated, and the solvating power of the polymerization medium may affect the propagation rate. Also an ion is always associated with a counter ion of opposite charge, which in the

Table 4.4 Ionic Initiators

Initiators	Sample Monomers[a]
Cationic types	
Lewis acids	
BF_3 (with H_2O, ROH, ROR)	1, 2
$AlCl_3$, $AlBr_3$ (with H_2O, ROH, RX)	2, 3
$SnCl_4$ (with H_2O)	3
$TiCl_4$	4
$FeCl_3$ (with HCl)	3
I_2	
Brønsted acids	
H_2SO_4	3(Low mol.wt.), 4
$KHSO_4$	
HF	3
$HClO_4$	
Cl_3COOH	5
Active salts	
$(C_6H_5)_3C^+BF_4^-$; $(C_6H_5)_3C^+SbCl_6^-$	3
$C_2H_5O^+BF_4^-$	3, 4
$Ti(OR)_4$	

Anionic types

Free metals[b]   in ⬡CH₃ , ⬡⬡ , liq.NH₃

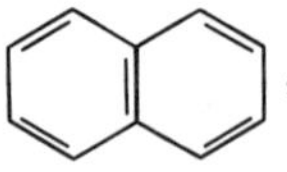

Na

K   etc.   3, 6, 7

Bases and salts

KNH_2, $NaNH_2$	3, 6
$Ar_2N^-K^+$	8
NaCN	
$NaOCH_3$	3
RLi, RK, RNa	9

$\left[\begin{array}{l} \text{R may be: } C_4H_9^-, \quad CH_2^- \\ \phi_3C^-, \phi CH_2^-, \phi\text{---}\diagup \end{array} \right]$

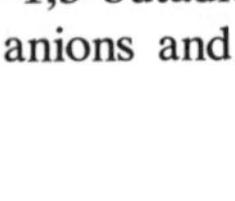

[a] 1 = 2-butene; 2 = isobutene; 3 = styrene; 4 = propylene; 5 = isopropenylben-zene; 6 = butadiene; 7 = stilbene; 8 = 2-cyano-1,3-butadiene; 9 = acrylonitrile.
[b] Operate by way of production of radical anions and subsequent reaction of these to actual initiating anions.

Table 4.5 Methods of Polymerizing Monomers

Monomer		Anionic	Cationic	Free Radical	Metal Oxide or Coordination Catalyst
$CH_2=CH_2$	Ethylene		+	+	+
$CH_2=CHCH_3$	Propylene		+		+
$CH_2C(CH_3)_2$	Isobutene		+		
$CH_2=CH—CH=CH_2$	Butadiene	+		+	+
$CH_2=C(CH_3)CH=CH_2$	Isoprene	+		+	+
$CH_2=CHC_6H_5$	Styrene	+	+	+	+
$CH_2=CHNO_2$	Nitroethylene	+			
$CH_2=CHOR$	Vinyl ethers		+		+
$CH_2=CH—N{<}^{CO—CH_2}_{CH_2—CH_2}$	Vinyl pyrrolidone		+	+	
$CH_2=C(CH_3)COOCH_3$	Methyl methacrylate	+		+	+
$CH_2=C(CN)COOCH_3$	Methyl α-cyanoacrylate	+		+	
$CH_2=CHCN$	Acrylonitrile	+		+	

foregoing equation is Li^+. This counter ion may be completely dissociated from the negative ion or it may be associated with it as an ion pair, and this too can affect the course of propagation.

$$\sim\sim\sim CH_2CH^-Li^+ + CH_2=CH \longrightarrow \sim\sim\sim CH_2CHCH_2CH^-Li^+$$
$$\quad\quad\; | \quad\quad\quad\quad\; | \quad\quad\quad\quad\quad\quad\; | \quad\quad | $$
$$\quad\quad X \quad\quad\quad\quad X \quad\quad\quad\quad\quad\quad X \quad\; X$$

The use of anionic initiation leads to a radical ion that can propagate at both ends of the polymer chain. If styrene is treated with sodium in naphthalene the sodium first transfers an electron to the naphthalene, which in turn transfers it to the styrene. The styrene has become an anion with an odd number of electrons, that is, it is also a free radical and is called a radical ion. It will combine with more monomer to give a chain with an anionic end and a free radical end.

This species could conceivably add monomer from the two ends by different mechanisms. More likely, because the charge is sufficiently delocalized, two of the radical ends couple to give a divalent anion that

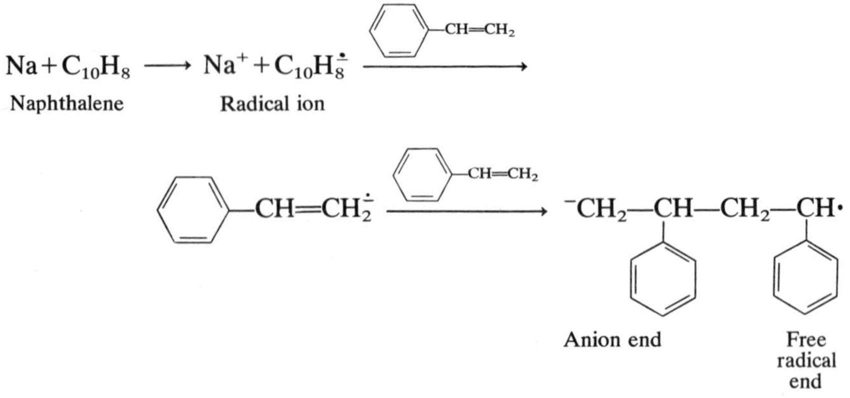

$$Na + C_{10}H_8 \longrightarrow Na^+ + C_{10}H_8^{\cdot -} \xrightarrow{\quad} $$

Naphthalene   Radical ion

-CH=CH$_2$

$$\text{Anion end} \qquad \text{Free radical end}$$

$$^-CH_2{-}CH{-}CH_2{-}CH{\cdot}$$

propagates from both ends by an ionic mechanism:

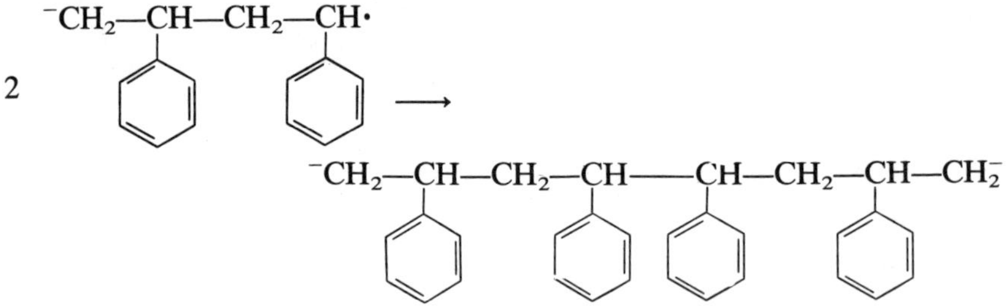

$$2 \quad {}^-CH_2{-}CH{-}CH_2{-}CH{\cdot} \longrightarrow$$

$$^-CH_2{-}CH{-}CH_2{-}CH{-\!-}CH{-}CH_2{-}CH{-}CH_2^-$$

Chain termination is more complicated than in free radical polymerization where it takes place by way of coupling and disproportionation (Sec. 4.3.1). Neither of these is possible because two negative ions cannot easily come together. The reluctance of ionic chains to terminate leads to the so-called "living" polymers (Sec. 4.3.7). Termination may result because of proton transfer from solvent, weak acid, polymer, or monomer. Thus water will quench an anionically initiated polymer. Proton transfer is not true destruction of transient species, and termination only occurs if the new species is too weak to propagate.

$$\text{\small\textasciitilde\textasciitilde\textasciitilde}CH_2{-}\underset{\underset{X}{|}}{CH}{}^- \xrightarrow{H_2O} \text{\small\textasciitilde\textasciitilde\textasciitilde}CH{=}\underset{\underset{X}{|}}{CH} + OH^-$$

Termination·

The recombination of a chain with its counter ion or the transfer of a hydrogen to give terminal unsaturation, frequent in cationic polymerization (see below), is unlikely in anionic systems. For example, if the counter ion is Na$^+$ the transfer to it of H$^-$ is improbable.

Termination can also be brought about by a cation-generating small molecule such as silicon tetrachloride. Four chains can terminate at the silicon atom so the molecular weight of the polymer has been quadrupled.

$$4 \sim\sim\sim CH_2-\underset{\underset{X}{|}}{CH}^-Li^+ \xrightarrow{SiCl_4} \left(\sim\sim\sim CH_2-\underset{\underset{X}{|}}{CH} \right)_4 Si + 4LiCl$$

Termination

For a "three-armed star-shaped" polymer a terminating agent such as 1,3,5-tris(chloromethyl) benzene may be used. This is a unique aspect of ionic termination that has no counterpart with free radicals. The radial block polymer is much less viscous than a linear polymer of similar molecular weight, and it is more soluble simply because its shape provides more opportunities for solvation. Thus it couples the benefits of very high molecular weight with easier handling properties.

An alternative procedure for preparing star-shaped polymer is to start with multifunctional initiators such as $C(CH_2C_6H_4Li)_4$. An initiator for a three-armed star-shaped polymer is the alkoxide of triethanolamine, $N(CH_2CH_2ONa)_3$.

Ionic polymerization may also be cationic. Table 4.4 shows which monomers may be polymerized cationically. Initiation is by proton donors such as conventional acids and Lewis acids, and these give rise to carbonium ions. Boron trifluoride in water is typical of a Lewis acid.

$$BF_3 + H_2O \longrightarrow H^+(BF_3OH)^-$$

$$H^+(BF_3OH)^- + CH_2=\underset{\underset{X}{|}}{CH} \longrightarrow CH_3\underset{\underset{X}{|}}{CH}{}^+(BF_3OH)^-$$

Initiation

$$\sim\sim\sim CH_2\underset{\underset{X}{|}}{CH}{}^+(BF_3OH)^- + CH_2=\underset{\underset{X}{|}}{CH} \longrightarrow \sim\sim\sim CH_2\underset{\underset{X}{|}}{CH}CH_2\underset{\underset{X}{|}}{CH}{}^+(BF_3OH)^-$$

Propagation

$$\sim\sim\sim CH_2\underset{\underset{X}{|}}{CH}{}^+(BF_3OH)^- \longrightarrow \sim\sim\sim CH=\underset{\underset{X}{|}}{CH} + H^+(BF_3OH)^-$$

$$\sim\sim\sim CH_2-\underset{\underset{X}{|}}{CH}{}^+(BF_3OH)^- \xrightarrow{NH_3} \sim\sim\sim CH_2-\underset{\underset{X}{|}}{CH}NH_2 + H^+(BF_3OH)^-$$

Termination

Propagation occurs as in anionic polymerization, and termination occurs when a proton is transferred back to the counter ion leaving a polymer molecule with terminal unsaturation. Unlike anionic polymerization the initiator is regenerated and can go on to generate other chains or even to attack the solvent. Termination can also be brought about by addition of a small molecule such as ammonia, and a polymer with an amine end group is formed. Again the initiator is regenerated.

It is only by way of ionic polymerization that functional end groups can be attached to polymer molecules. With anionic polymer molecules that grow in two directions, as described, CO_2 yields carboxyl groups; ammonia will provide amine end groups; potassium isocyanate, isocyanate end groups; and HCl, chlorine end groups. If the molecular weight of a polymer is very high the effect of these end groups is negligible. On the other hand, if the ratio of initiator to monomer is such that the number of recurring units is low, an oligomer is obtained. This termination procedure provides an elegant method for the manufacture of difunctional compounds such as dibasic acids and diisocyanates. But, as so often happens, elegance and expense go together. One mole of expensive catalyst is required for every mole of difunctional compound produced.

4.3.7 LIVING POLYMERS

Living polymers are an important ramification of ionic polymerization. The polymer theory that has been outlined so far was developed between 1935 and 1950. The importance of initiation and propagation reactions was recognized, but no one worried very much about termination. It was understood that termination was more difficult in ionic than free radical polymerization because the charges on the growing chains repelled one another, but it was assumed that termination would come about somehow or other. Eventually it was realized that it did not have to occur at all. If styrene is polymerized ionically with sodium naphthalide in tetrahydrofuran solution and care is taken not to introduce agents that terminate chains, a polymer is formed whose chain length can be estimated from the viscosity of the polystyrene solution. The ends of the chains are unterminated, and the polymer is described as "living." If further styrene is added weeks or even months later there will be a marked increase in viscosity showing that the polymer chains have started to grow again once they are supplied with fresh monomer.

4.3.8 BLOCK COPOLYMERS

Instead of styrene in the above example some other monomer such as isoprene may be added to the living polymer, and a copolymer results. Copolymers can be achieved by other means (Sec. 4.3.3), but they are

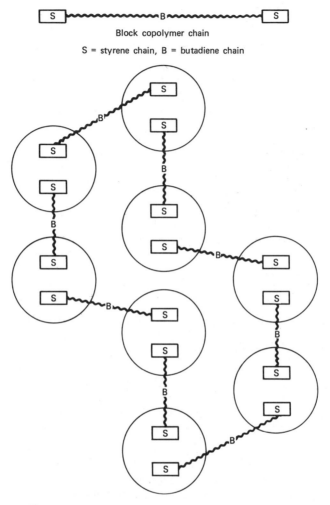

Figure 4.5 Block copolymer: styrene-butadiene-styrene.

usually random. With the living polymer technique the copolymer is ordered, consisting of a chain of X molecules followed by a chain of Y molecules. If desired a further set of X molecules or of any other monomer can be added. Such materials are called block copolymers.

Block copolymers can also be made by condensation techniques as described for "Spandex" formation (Part II, Sec. 3.10). Condensation reactions can be used to prepare block polymers only if the reactions take place at low temperatures. At high temperatures existing bonds in many polymers break and reform to provide random distribution and thus a random copolymer.

Block copolymers have industrial applications especially in the formation of elastomers. They may be constructed in such a way that one block plasticizes the other. Conversely the blocks may be insoluble in one another and tend to repel each other. They will however tend to associate with similar blocks in other polymer molecules. This is illustrated in Figure 4.5 for a styrene-butadiene-styrene block copolymer. It is quite different from the common free radical polymerized styrene-butadiene rubber, which is a random copolymer. Polybutadiene is a flexible rubbery material; polystyrene is hard and brittle. Furthermore polystyrene is highly insoluble in polybutadiene. The polystyrene ends therefore associate with other polystyrene ends with which they are more compatible, and physical bonding, that is, Van der Waals forces, results. Although these forces are not very strong they nonetheless give an element of cross-linking so that the polymer at room temperature has many of the properties of a cross-linked material. At higher temperatures, however, the weakness of the forces dissociates the "cross-links" so that the polymer can be processed as if it were a simple thermoplastic.

The useful range of temperature for a block copolymer is determined by the glass transition temperatures (T_g) of the blocks that constitute it. The glass transition temperature is a property of amorphous polymers and is discussed in more detail in Section 4.5.2. At this stage it is sufficient to say that if we melt an amorphous polymer and then allow it to cool it will at its precise glass transition temperature cease to be soft, pliable, flexible, and plasticizing and become hard, rigid, and glassy.

If the styrene-butadiene-styrene block copolymer is to be an elastomer it must be used above the T_g for polybutadiene. Equally, if it is to retain physical cross-linking it must be used below the T_g for polystyrene.

Polymers of this sort have found application in rubber footwear, in rubber soles for shoes, and in both solvent-based and hot melt adhesives.

4.3.9 GRAFT COPOLYMERS

Chains are usually grafted onto a polymer backbone by creation of a free radical site along the backbone, which initiates growth of a polymer chain. Less often the backbone possesses functional groups, and chains can be condensed onto it.

An example of graft copolymerization is the production of high impact polystyrene. Polystyrene is a useful low cost plastic. Unfortunately it is brittle, and under stress it tends to craze or stress crack. These defects are alleviated by graft copolymerization. Polybutadiene is dissolved to the extent of 5–10% in monomeric styrene and an initiator added. Because polybutadiene readily undergoes chain transfer, polystyrene chains grow on the polybutadiene backbone, and an impact resistant graft copolymer results.

Polyacrylonitrile chains can be grafted onto a starch backbone with the aid of a ceric sulfate initiator or ionizing radiation from a cobalt 60 source. Typically three chains of acrylonitrile, each with a molecular weight of about 800,000, graft onto each starch molecule. The graft copolymer has markedly different properties from starch itself, able to absorb as much as 1000 times its own weight of water.

4.3.10 METAL COMPLEX CATALYSTS

The third method to bring about chain growth polymerization is by the use of metal complex catalysts. Karl Ziegler, who spent World War II at the Kaiser Wilhelm Institute in Germany trying to find ways to polymerize small molecules into gasoline, found that titanium tetrachloride or titanium trichloride combined with an aluminum alkyl catalyzes the polymerization of ethylene. The two components of the catalyst form a solid complex, a proposed structure for which is shown on p. 198. Ziegler found that his catalyst produced a high molecular weight linear crystalline polyethylene without any of the chain branching or oxygen bridges obtained in the high pressure free radical polymerization. It was stronger and denser than the conventional material. The conditions required—about atmospheric pressure and 60°C—were astonishingly mild. Ziegler offered his discovery to Imperial Chemical Industries in the United Kingdom at a remarkably low price, but the latter were heavily committed to their own high pressure process and were not interested. Other companies did license his process throughout the world, but a competing process developed by Phillips Petroleum in the United States proved to have advantages (Sec. 4.3.11) and is now used more widely. Thus Ziegler's contribution, which attracted well-earned attention throughout the scientific world, did not find its greatest application in polyethylene manufacture.

In 1955, about three years after Ziegler's breakthrough, the Italian chemist, Giulio Natta, who was working for the Italian chemical giant, Montecatini (now Montedison), tried the new catalyst system on propylene. It does not take great scientific intuition to realize that if a catalyst works on ethylene it might also work on propylene. This however had not been the case with free radical polymerization. The allylic hydrogens on propylene were labile and easily displaced, so that several free radical sites developed on the monomer and growing polymer. A useless, low

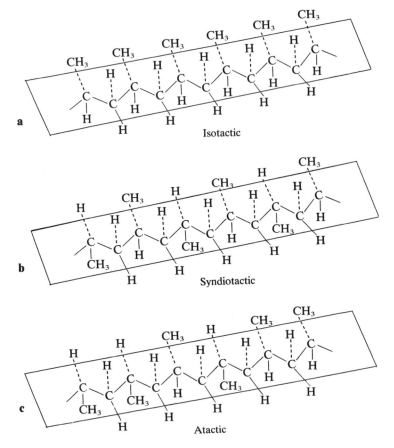

Figure 4.6   Stereoregular polymers.

molecular weight, cross-linked polymer was obtained. With Ziegler catalysis, however, the propylene polymerized smoothly to a high molecular weight linear polymer, and in addition—and this was the dramatic thing that won both Ziegler and Natta a Nobel prize—the polymer was stereoregular. Few discoveries in organic chemistry have created as much interest or excitement.

A stereoregular polymer may be defined as one in which the substituent groups are oriented regularly in space. Structure **a** of Figure 4.6 shows such a polymer with a substituent, CH_3, in a regular formation. The carbon atom on which the substituent occurs is asymmetric, that is, four different groups are attached to it. It had long been recognized that polymers could contain asymmetric carbon atoms, but Natta was the first person to synthesize such a polymer in which all the asymmetric carbons had the same orientation. This polymer is said to be isotactic because all the substituents are similarly placed. Another type of stereoregularity is shown in **b** of Figure 4.6 in which the substituents point alternately forward and back. Such polymers are said to be syndiotactic. Low temperature tends to favor formation of syndiotactic structures. The conventional nonstereoregular polymers (**c**, Fig. 4.6) are said to be atactic and have their substituents placed randomly.

The discovery of Ziegler-Natta catalysis meant that almost overnight a procedure had become available for polymerization of unsaturated compounds that could not be polymerized by way of free radicals. The method was versatile and offered scope for further research. There are many transition metal salts besides the titanium trichloride and tetrachloride that Ziegler used and an equally large number of organometallic compounds with which to combine them. The adding of ligands increased the possibilities for influencing results. Since Ziegler's discovery hundreds of combinations and thousands of ratios of constituents have been evaluated. The polymer chemist today has power to achieve practically any molecular configuration he thinks will give him the properties he is seeking. It is this versatility that makes Ziegler-Natta catalysis such a powerful tool. Metal oxide catalysis (Sec. 4.3.11), so important for ethylene polymerization, does not have this versatility.

This power is illustrated by the polymerization of butadiene. Polybutadiene may have either a 1,2 or a 1,4 configuration (Fig. 4.7). The 1,4 polymer has a double bond, and the chain structure can be *cis* or *trans*. The 1,2 polymer has vinyl side chains, and these can be arranged in atactic, isotactic, or syndiotactic configurations. Thus five different polybutadienes exist, and all of them have been synthesized with the aid of Ziegler-Natta catalysts. Figure 4.7 shows the various catalyst combinations. These were discovered by trial and error. *Cis*-1,4-butadiene is an elastomer resembling natural rubber, whereas a less regular polymer is harder and stiffer and may be used for can coatings.

The stereospecificity of Ziegler catalysts has allowed chemists to do what nature with its highly specific enzymes has been doing. Because of these enzymes nature is able to synthesize optically active compounds, sterically complex antibiotic molecules, and also stereoregular polymers. Natural *hevea* rubber is *cis*-1,4-polyisoprene while *trans*-1,4-polyisoprene is the nonelastomeric *balata* or *guttapercha*. In the synthesis of *hevea* rubber in nature, every step is catalyzed by an enzyme. The starting material is acetic acid, a material manipulated with particular ease by nature as is illustrated by its various metabolic pathways in the

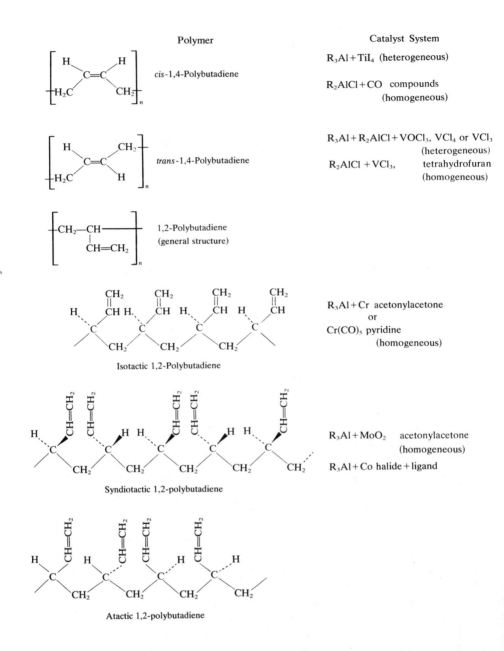

Figure 4.7 Catalyst systems for different polybutadienes.

body. The final step is the polymerization of the monomer, isopentenyl pyrophosphate, which nature prefers over isoprene, by a polymerase present in the rubber plant.

$$CH_3COOH \xrightarrow{-H_2O} CH_3CCH_2COOH \longrightarrow CH_3CCH_2COOH$$

with the intermediate structures showing:

$$CH_3\text{C}(=O)CH_2COOH$$

$$\xrightarrow[-CO_2]{-H_2O} \underset{CH_3}{\overset{CH_3}{>}}C=CH-COOH \xrightarrow{\text{Reduction}} \underset{CH_3}{\overset{CH_3}{>}}C=CH-CH_2OH \longrightarrow$$

$$\underset{H_2C}{\overset{CH_3}{>}}C-CH_2-CH_2OH \xrightarrow{H_4P_2O_7} \underset{H_2C}{\overset{CH_3}{>}}C-CH_2 \quad CH_2-O-\overset{OH}{\underset{O}{P}}-O-\overset{OH}{\underset{O}{P}}-OH$$

Isopentenylpyrophosphate

$$\xrightarrow[\substack{Hevea \\ brasiliensis}]{Enzyme \\ in} \left[\underset{CH_2}{\overset{CH_3}{>}}C=CH \atop CH_2 \right]_n + HO-\overset{OH}{\underset{O}{P}}-O-\overset{OH}{\underset{O}{P}}-OH$$

cis-1,4-Polyisoprene   Pyrophosphoric acid

With Ziegler catalysis scientists can duplicate nature's precision and produce materials that are similar to either *hevea* rubber or *gut-tapercha*. Thus the chemical industry can and does produce "synthetic natural rubber." It is not quite the same as natural rubber because both end groups and molecular weight distribution differ. Hence it is almost but not quite as resilient. Equally significant on the plane of ideas is that Ziegler catalysis enables us to mimic nature by making stereoregular polymers. It is satisfying too that Ziegler catalysis works, as does nature, at moderate temperatures and pressures compared with, for example, the formidable 1200 atm and 200°C required for traditional low density polyethylene synthesis.

What causes stereospecificity? How do Ziegler catalysts work? The mechanism is not yet fully understood, but it is known that polymerization takes place at active sites on the catalyst surface. The catalyst is an electron deficient solid complex of an aluminum alkyl and a titanium halide, the alkyl group on the titanium atom coming from the alkyl aluminum portion of the catalyst. This is one of several possible structures.

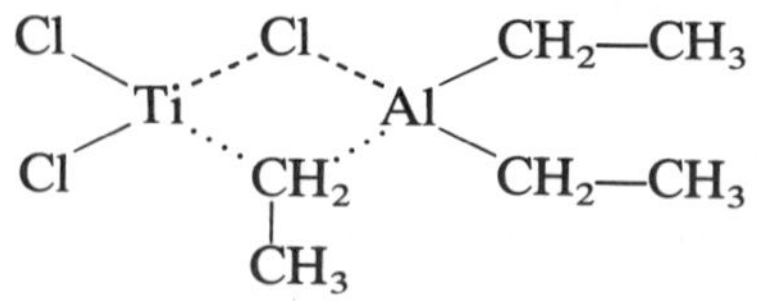

The electron deficiency occurs between the titanium-carbon and carbon-aluminum bonds. Titanium has an octahedral configuration with one ligand vacancy, as shown in Figure 4.8, and a monomer, for example propylene, may become pi-bonded to the titanium. After this insertion new bonds form and the old ones break. The net result is that the propylene molecule inserts itself between the titanium atom and the alkyl group. The ligand vacancy again exists so that the same progression can happen all over again, and another propylene molecule can be incorporated into the alkyl chain. This is the propagation step, and the polymer chain grows by successive insertion of monomer units at the surface of the titanium complex catalyst.

The system is heterogeneous, and the catalyst is insoluble in the monomer and in the solvent. The insertion of monomer molecules takes place at the solid-liquid interface, and the polymer chain grows from the insoluble catalyst into the solvent. It is the solvating effect of the solvent on the polymer that attracts the chain away from the catalyst surface and into the solvent and allows further monomer to have access to the titanium atom. Some homogeneous metal-catalyzed polymerizations can also be carried out.

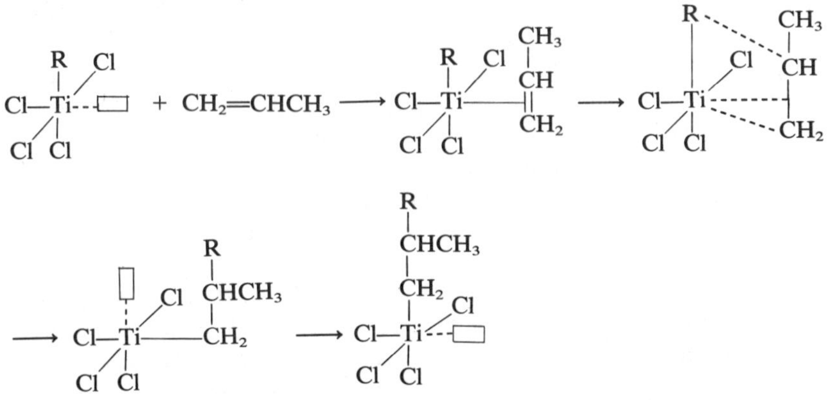

Figure 4.8 Mechanism of Ziegler-Natta polymerization.

The mild conditions in the use of Ziegler-Natta catalysis are not only a bonus but also a prerequisite for it. At higher temperatures bonds around the catalyst would tend to break and reform, and stereospecificity would be lost. The mild conditions also insure linear polymers and eliminate the branching that is characteristic of free radical initiation. Furthermore, the linear chains can get very near to each other, which gives them high cohesive strength and crystallinity and confers certain desirable properties discussed in Section 4.5.1.

It should be noted that stereoregularity does not always depend on the presence of an asymmetric carbon atom. Double bonds in the polymer chain as in polybutadiene or polyisoprene provide the basis for polymers with *cis* and *trans* structures in the recurring polymer unit.

4.3.11 METAL OXIDE CATALYSTS

At the same time that Ziegler was working in Germany, studies in the United States were underway using supported metal oxide catalysts. Researchers for Standard Oil of Indiana developed a molybdenum oxide catalyst supported on silica or aluminum that gives high density polyethylene. Their discovery predated Ziegler's but they did not exploit it because a consultant's evaluation was negative. The consultant, interested in making film, could see no virtue in a stiff, structural-like polymer. The conventional wisdom associated with this story is that it is not enough to invent. One must also recognize the importance of the invention. Standard Oil of Indiana (AMOCO) does use the process now.

Another oxide catalyst system, chromic oxide supported on silica or alumina, was developed almost concurrently with Ziegler's catalyst by Phillips Petroleum. With it polymers can be obtained of higher molecular weight than those obtained by the Ziegler method. Indeed, it is necessary to include so-called chain stoppers, monofunctional reactants that behave like chain transfer agents (Sec. 4.3.2), and prevent formation of a very high molecular weight, intractable polymer. Furthermore, the polymerization is carried out in such a way that the polymer precipitates as a fine powder. The Ziegler process provides a molten mass that can be handled only by extruding it as "ropes" and chopping the ropes into pellets. Phillips licensed their process widely, and it is today the preferred way to make linear polyethylene.

The mechanism of metal oxide catalysis is uncertain. Theoretical interest, perhaps unjustifiably, is less because neither the Standard Oil of Indiana nor the Phillips process can be used to make polypropylene. It would be of interest to understand enough about catalyst theory to know why.

4.4 POLYMERIZATION OF THERMOSETS

Although the mechanism of polymerization processes is frequently complex, the overall process can usually be written simply. In the case of bifunctional monomers, there is either chain polymerization, $nCH_2\!=\!CHX \rightarrow \{CH_2\!-\!CHX\}_n$, or step growth polymerization as shown in Section 4.1. Sometimes only one monomer is required, as with caprolactam.

Polymerizations with monomers of functionalities greater than 2 are more difficult to visualize because the products are three dimensional. The condensation of a difunctional and trifunctional reagent is shown schematically in Figure 4.3. An example is the unsaturated polyester resin (Sec. 4.1) where a linear polyester is first formed by step polymerization and is later cross-linked by chain polymerization.

This technique is convenient but others are possible. With phenol formaldehyde resins (see below) both stages are step growth, and with ethylene-propylene rubbers (Part II, Sec. 4.2.3) both are chain growth.

The description of the production of three groups of polymers involving monomers with functionalities greater than 2 and that are consequently thermosets follows.

4.4.1 PHENOPLASTS AND AMINOPLASTS

Phenol-formaldehyde resins (phenoplasts) may be prepared from phenol and formaldehyde in two ways, both involving step polymerization. The first method employs an alkaline catalyst and excess formaldehyde and gives a cross-linked structure in one operation (a "one-stage" resin). The methylolphenols condense to give low molecular weight linear polymers called resoles which contain occasional oxygen bridges. Typical structures are:

Saligenin   Homosaligenin

On further heating the free methylol groups condense to give a cross-linked polymer.

The second method uses an acid catalyst and excess phenol to give a linear thermoplastic resin that may be stored or sold in that form. The linear polymers are called novolacs, and, unlike resoles, they have no free methylol groups and thus cannot cross-link. Treatment with more formaldehyde and alkali leads to formation of free methylol groups, and cross-linking can occur. The reagent is usually hexamethylene tetramine, which on heating generates formaldehyde as reactant and ammonia which provides the alkaline conditions.

Typical novolac structure

In these reactions the formaldehyde shows a functionality of 2 while the phenol has three active sites—the two positions ortho and the one para to the hydroxyl group. Only two of them are used in novolac formation because of the scarcity of formaldehyde. Thermoplastic phenolics useful for the preparation of varnishes result from the condensation of formaldehyde with p-substituted phenols.

Urea will also give cross-linked resins with formaldehyde (aminoplasts) under slightly alkaline conditions. Methylolureas are formed first. There follows a series of condensations that include ring formation, and the product is a complex thermoset polymer of poorly defined structure of which the following may be typical.

Crosslinked urea formaldehyde

Urea has a functionality of 4 in urea-formaldehyde resins, corresponding to the four labile hydrogen atoms. Melamine (Sec. 2.12.1) with three amino groups and six labile hydrogens has a functionality of 6, and it too will form thermoset resins with formaldehyde—the so-called melamine formaldehyde resins.

4.4.2 POLYURETHANES

If a diol undergoes step polymerization with a diisocyanate, a linear thermoplastic polyurethane is obtained because both monomers are bifunctional.

2:6 Tolylene diisocyanate
(A mixture of the 2:4 and 2:6 isomers is usually used.)

Low molecular weight poly(propylene glycol)

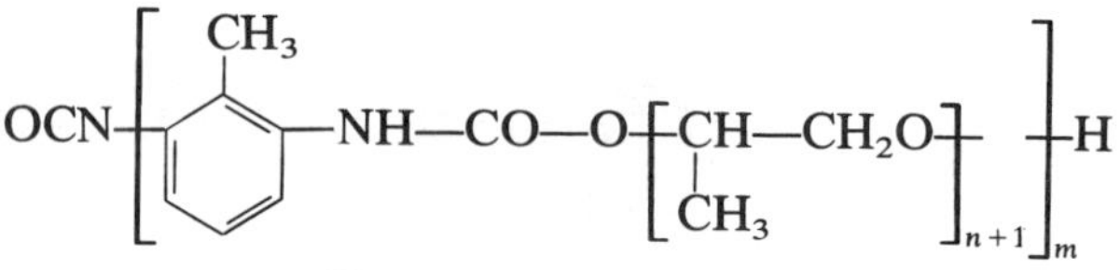

$$OCN \left[\underset{CH_3}{\underset{|}{\bigcirc}} -NH-CO-O \left[\underset{CH_3}{\underset{|}{CH}}-CH_2O \right]_{n+1} \right]_m H$$

Thermoplastic polyurethane

The poly(propylene glycol) is synthesized from propylene oxide as follows:

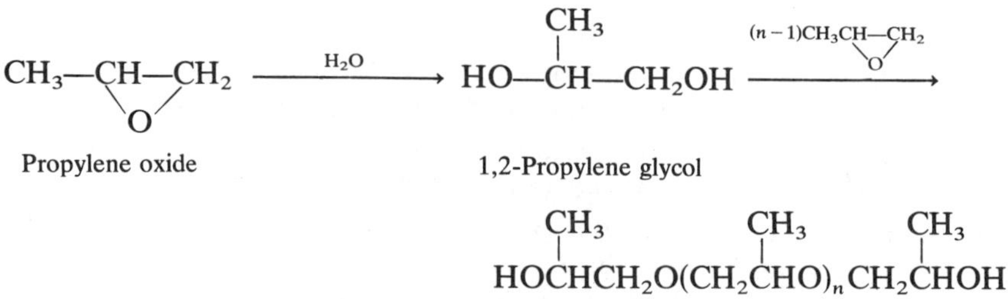

$$CH_3-\underset{O}{\underset{\diagdown\diagup}{CH-CH_2}} \xrightarrow{H_2O} HO-\underset{CH_3}{\underset{|}{CH}}-CH_2OH \xrightarrow{(n-1)CH_3CH-CH_2}$$

Propylene oxide   1,2-Propylene glycol

$$HO\underset{CH_3}{\underset{|}{CH}}CH_2O(CH_2\underset{CH_3}{\underset{|}{CH}}O)_n CH_2\underset{CH_3}{\underset{|}{CH}}OH$$

Poly(propylene glycol)

To obtain the more useful cross-linked polyurethanes, trifunctional reagents are required. Sometimes tolylene diisocyanate (TDI) is converted to a trifunctional reactant by reaction with trimethylolpropane. The new reagent has the advantage of being considerably less toxic than TDI. Alternatively, trifunctional hydroxyl compounds may be made by reaction of propylene oxide with glycerol.

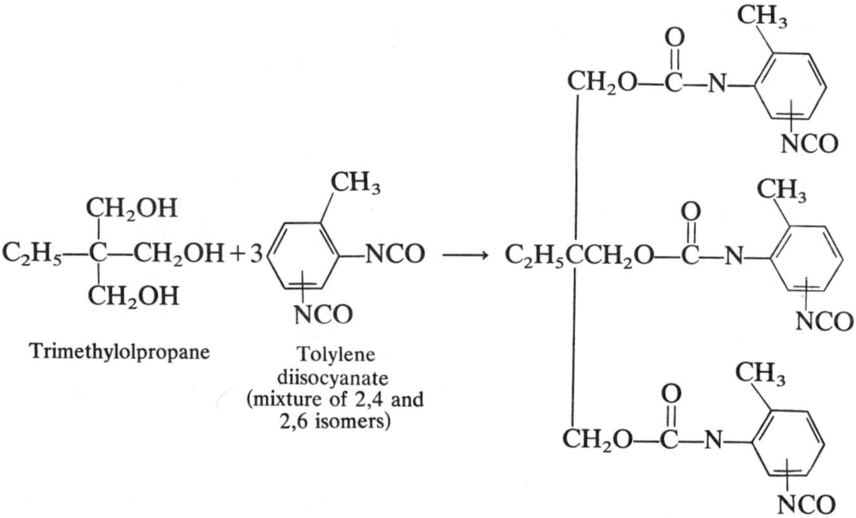

Trimethylolpropane   Tolylene
diisocyanate
(mixture of 2,4 and
2,6 isomers)

$$CH_2O-(CH_2-\overset{\overset{\displaystyle CH_3}{|}}{C}HO)_n-CH_2\overset{\overset{\displaystyle OH}{|}}{C}HCH_3$$

$$\begin{array}{l} CH_2OH \\ CHOH+3n CH_3CH-CH_2 \overset{KOH}{\longrightarrow} \\ CH_2OH \qquad\qquad\quad O \end{array}$$

$$CHO-(CH_2-\overset{\overset{\displaystyle CH_3}{|}}{C}HO)_n-CH_2\overset{\overset{\displaystyle OH}{|}}{C}HCH_3$$

$$CHO-(CH_2-\overset{\overset{\displaystyle CH_3}{|}}{C}HO)_n-CH_2\overset{\overset{\displaystyle OH}{|}}{C}HCH_3$$

Glycerol   Proplyene oxide

Castor oil (Sec. 3.2.8) is a naturally occurring triglyceride that contains three hydroxyl groups and is another useful starting material.

TDI is the raw material for about 60% of polyurethanes. MDI which is 4,4′-diphenylmethane diisocyanate and oligomers of it is also important. The presence of the trimer and tetramer in the product mixture means that the product has a functionality greater than 2. The aliphatic isocyanate HMDI, hexamethylene diisocyanate, $OCN-(CH_2)_6-NCO$, leads to coatings with good color retention and weathering properties. It is toxic and is used in the form of a trimer with a biuret structure.

A biuret is formed by the interaction of 1 mole of an isocyanate with 1 mole of a urea. In the case of HMDI biuret, 2 moles of diisocyanate react with water to give a urea with the release of 1 mole of CO_2. The third mole of diisocyanate reacts with the urea, and the trimer has three isocyanate groups:

$$3OCN-(CH_2)_6-NCO \xrightarrow[-CO_2]{+H_2O} R-N \begin{array}{l} \overset{\overset{\displaystyle O}{\|}}{C}-NHR \\ \\ \overset{\displaystyle C}{\underset{\displaystyle O}{\|}}-NHR \end{array} \qquad [R=OCN-(CH_2)_6-]$$

Hexamethylene diisocyanate

Biuret trimer
"desmodur N"

A third linkage found in foams is the allophonate group, which forms from the interaction of an isocyanate with a urethane linkage.

$$\sim\sim\sim NH\overset{\overset{\displaystyle O}{\|}}{C}O\sim\sim\sim + OCN\sim\sim\sim \longrightarrow \sim\sim\sim N\overset{\overset{\displaystyle O}{\|}}{C}O\sim\sim\sim$$

Urethane   Isocyanate

$$\begin{array}{c} | \\ C=O \\ | \\ \sim\sim\sim NH \end{array}$$ Allophonate

Still another form in which isocyanates are used is as isocyanurates. These are isocyanate trimers and have an advantage over biurets in being more stable. A typical isocyanurate is the trimer of 3-isocyanatomethyl-3,5′,5-trimethylcyclohexyl isocyanate, trivially known as "isophorone diisocyanate." The conversion to trimer takes place in the presence of a basic catalyst.

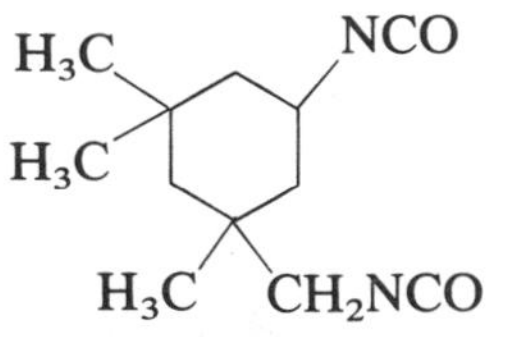

3-Isocyanatomethyl-3,5′,5-trimethylcyclohexyl isocyanate, "Isophorone diisocyanate"

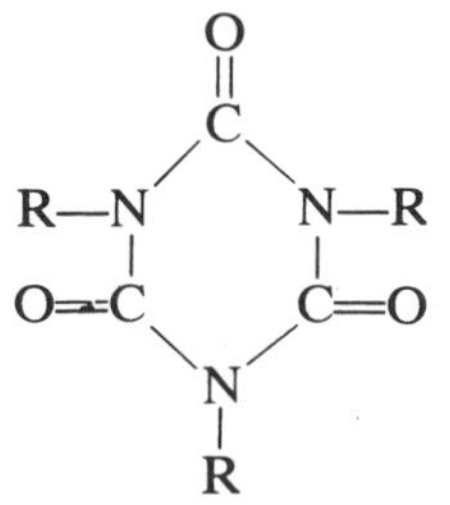

Isocyanurate trimer of isophorone diisocyanate

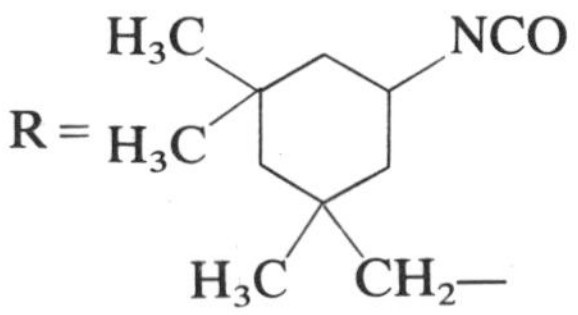

"Isophorone diisocyanate" is derived from isophorone (Fig. 2.14) as follows:

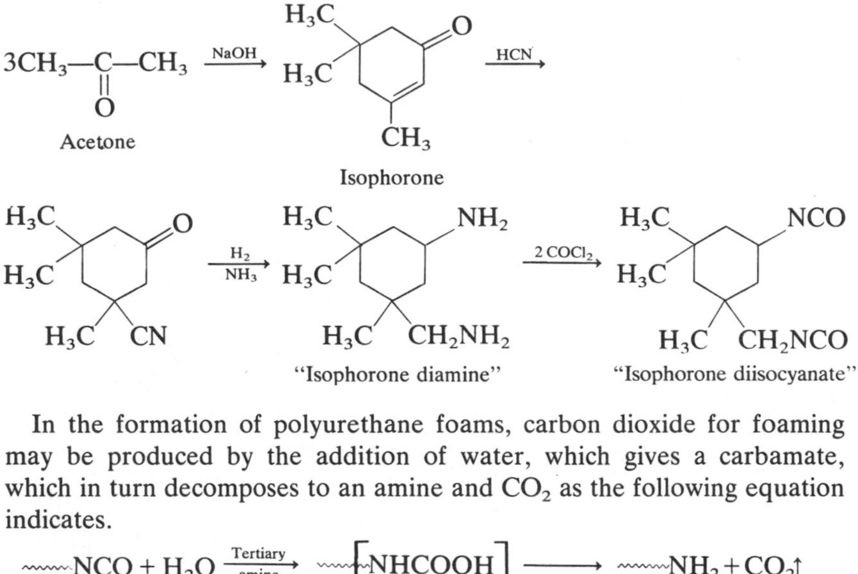

Acetone

Isophorone

"Isophorone diamine"

"Isophorone diisocyanate"

In the formation of polyurethane foams, carbon dioxide for foaming may be produced by the addition of water, which gives a carbamate, which in turn decomposes to an amine and CO_2 as the following equation indicates.

$$\text{~~NCO} + H_2O \xrightarrow[\text{amine}]{\text{Tertiary}} \text{~~}\left[\text{NHCOOH}\right] \longrightarrow \text{~~NH}_2 + CO_2\uparrow$$

Carbamate

Amine   Carbon dioxide

This, however, is an expensive way to obtain a gas for foaming, and accordingly fluorocarbons are also used as foaming agents. When water is added urea linkages form by the condensation of the amine with more isocyanate as the following equation shows:

$$\text{wwwNCO} + \text{NH}_2\text{www} \longrightarrow \text{wwwNHC—NHwww}$$
$$\underset{\displaystyle O}{\overset{\displaystyle \|}{}}$$

Polyurethane foams contain both these and biuret linkages.

4.4.3 EPOXY RESINS

The curing of epoxy resins is interesting because functionality is generated in the course of the reaction.

Epoxy resins are typically condensates of bisphenol A with epichlorohydrin. If a large excess of epichlorohydrin is used a simple molecule results from the condensation of two moles of epichlorohydrin with one mole of bisphenol A.

Epichlorohydrin   Sodium salt of bisphenol A   Epichlorohydrin

If the reactants are close to equimolar, on the other hand, a low molecular weight polymer is formed where n is between 1 and 4. In either case the terminal groups are epoxy groups.

The epoxy groups will react with a multifunctional amine such as ethylene diamine to give a cross-linked resin. When one of the primary amine groups reacts with an epoxy group a hydroxyl group and a secondary amine group are generated. Both of these groups can react further with epoxy groups in principle, although the hydroxyl group will react only at high temperatures. The aliphatic amine groups react at room temperature, and if two molecules of ethylene diamine react with one polymer molecule of epoxy resin a molecule is generated with four amine groups—two primary and two secondary. The new polymer has four amine groups with six active hydrogens, each of which can react with more epoxy resin, and thus the conditions for cross-linking have been established.

$$-NH_2 + -\overset{|}{\underset{\diagdown O \diagup}{C}}-\overset{|}{\underset{}{C}}- \longrightarrow -\overset{|}{\underset{OH}{C}}-\overset{|}{\underset{NH}{C}}-$$

$$H_2\overset{}{\underset{\diagdown O \diagup}{C}}-CH\sim \sim\sim CH-\overset{}{\underset{\diagdown O \diagup}{CH_2}} + 2H_2N-C_2H_4-NH_2 \longrightarrow$$

$$H_2N-C_2H_4-NH-CH_2-\overset{}{\underset{OH}{C}}\sim\sim\sim\sim\overset{}{\underset{OH}{C}}-CH_2-NH-C_2H_4-NH_2$$

The interaction of the secondary amine group with an epoxy group generates a tertiary amine. Tertiary amines are catalysts for the self-polymerization of epoxy groups to polyethers, so yet another polymerization mechanism has been introduced and is shown in the following equation. This polymerization is chain growth, whereas the polymer formation resulting from the condensation of amine and epoxy groups is step growth.

$$\sim\sim\sim\overset{}{\underset{\sim\sim}{N}}H + -\overset{|}{\underset{\diagdown O \diagup}{C}}-\overset{|}{\underset{}{C}}- \longrightarrow -\overset{|}{\underset{OH}{C}}-\overset{|}{\underset{N\underset{\sim\sim}{<}}{C}}-O-\sim\sim$$

Tertiary amine

$$-\overset{|}{\underset{\diagdown O \diagup}{C}}-\overset{|}{\underset{}{C}}- \xrightarrow{\text{Tertiary amine}} -[O-\overset{|}{\underset{|}{C}}-\overset{|}{\underset{|}{C}}]_n-$$

4.5 POLYMER PROPERTIES

In the previous section we described how monomers were converted to polymers, bifunctional monomers leading to linear thermoplastic resins and polyfunctional monomers providing cross-linked thermosetting resins. Only one property, average molecular weight, has been mentioned.

Molecular weight and mechanical strength are related since strength increases rapidly between 50 and 500 monomer units. Further increases in molecular weight have little effect. In this section we describe additional properties of polymers and discuss the factors that give them the properties that make them useful—high viscosity, tensile strength, and toughness.

4.5.1 CRYSTALLINITY

Crystallinity is the key factor governing polymer properties. The easiest way to think of it is to regard crystallinity as a situation in which the polymer chains fit into an imaginary pipe. That is, they align themselves in bundles with a high lateral order, and the chains lie side by side. To do this they must be linear, not coiled, and there must not be bulky groups or branching to prevent the polymer chains from achieving nearness.

There is an analogy, although not an exact one, with the crystallization of nonpolymeric materials like sodium chloride or n-hexane where the ions or molecules must fit into a crystal lattice. With polymers, however, it is long chain molecules, not ions or small molecules, that must fit, and the structure into which they fit is not a lattice but an imaginary cylindrical tube. Figure 4.9 attempts to illustrate this concept and shows how the chains may line up in an ordered fashion. It is, however, unlikely that all the chains in a polymer or even all of a single chain will be able to enter into the ordered structure of complete crystallinity, although nylon and high density polyethylene come close. Usually the ordered regions are small—microcrystalline—and are scattered through the polymer which is otherwise amorphous. Where the polymer is crystalline, it is platelike and of uniform thickness. Emanating from these crystalline regions are the sections of the chains that are not incorporated into the crystal lattice. They form the amorphous part of the polymer and may actually coil back over the crystalline platelets. Thus one can legitimately talk of the degree of crystallinity of a polymer. Some polymers are almost totally crystalline, others almost totally amorphous. Even a single polymeric substance can exist with a range of crystallinities depending on how it was made and processed.

Two factors govern the tendency of a polymer to crystallize. One is the ease with which the polymer chains will pack into a "crystal," and the other is the magnitude of the attractive forces between neighboring molecules of the polymer. The first of these means that crystalline polymers are more likely to form from chains that do not have bulky substituents and where there is not a great deal of branching. Poly(ethylene terephthalate) is without bulky side chains and is crystalline

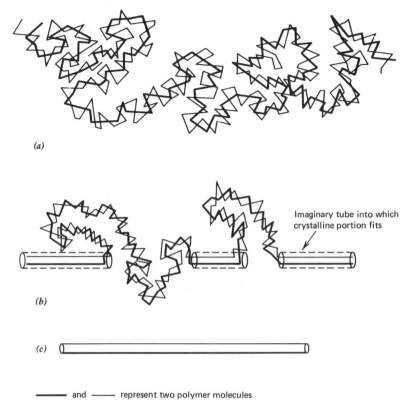

(a)

(b)

Imaginary tube into which
crystalline portion fits

(c)

———— and ——— represent two polymer molecules

Figure 4.9 Polymer crystallization. (*a*) Noncrystalline; (*b*) partially crystalline; (*c*) completely crystalline.

after orientation. The rigid benzene ring makes the polymer chains stiff
and unwilling to coil.

Isotactic polymers made by Ziegler-Natta polymerization are also
highly crystalline unless, as in *p*-substituted polystyrenes, a bulky group
keeps the polymer chains apart. Polystyrene polymerized as it normally is
by peroxide catalysis is an example of an amorphous polymer. Isotactic
polystyrene made by Ziegler-Natta catalysis, however, is a highly crystalline material with properties very much different from those of the
amorphous product. It is not an article of commerce because it crystallizes
very slowly and therefore changes its properties after processing.

Other amorphous polymers are exemplified by poly(methyl methacrylate) and polycarbonates (Part II, Sec. 2.1.8). The pendant groups on the
methacrylate polymer are bulky, and the two phenyl groups in the

polycarbonate are not coplanar. Thus crystal formation is obstructed. Most copolymers have little crystallinity because their structures are nonlinear.

In general crystalline polymers are opaque because light is reflected or scattered at the boundaries between the microcrystalline and amorphous regions. Amorphous polymers are transparent and glasslike. Two exceptions should be noted. If a crystalline polymer is biaxially oriented, for example in drawn poly(ethylene terephthalate) (Mylar) sheet, then the whole sheet is in effect a single crystal and is transparent. Furthermore, in a few polymers, the most important of which is poly(4-methylpentene), the refractive index of the crystal is identical with that of the amorphous region. No light scattering occurs at boundaries, and the crystalline polymer is clear and transparent.

The second factor leading to crystallization is the forces of attraction between neighboring molecules. These comprise hydrogen bonding, which is the strongest, and the various kinds of Van der Waals forces, dipole-dipole forces, dipole-induced dipole forces, and London dispersion forces. They vary in strength from about 1–2 kcal mole^{-1} (5–10 kJ mole^{-1}) per unit of polymer chain in elastomers to 5–10 kcal mole^{-1} (20–40 kJ mole^{-1}) in fibers. London forces are, weak whereas the others can be quite high. Cellulose and nylon provide examples of hydrogen bonding (Sec. 3.3.3; Part II, Sec. 3.7), PVC, and polyacrylonitrile of dipole-dipole interaction, and polyethylene of London forces.

We can extend the analogy between crystallization of n-hexane and polyethylene. As n-hexane is cooled the thermal motion of the molecules decreases until it can no longer overcome the forces of attraction between them. Accordingly the molecules pack into the orientation of lowest energy, that is, the crystal lattice, and the sample solidifies or crystallizes.

The molecules of polyethylene are hundreds of times larger than those of hexane. Although the intermolecular forces between the —CH$_2$— units are about the same in the two molecules the total force per molecule in the polymer will be much higher. Furthermore, the polyethylene chains will not have the freedom of movement of hexane molecules. Instead they will be wriggling and coiling. There is little chance that one of them will ever be fully extended. The chains will also become entangled with one another, which will hinder molecular motion. Molten polymers are viscous both because of chain entanglement and intermolecular forces. The latter factor is negligible at high temperatures, and viscosity under these conditions is due largely to chain entanglement.

When the temperature of molten polyethylene is reduced molecular motion diminishes as it does in hexane. Eventually there will be a tendency for the chains to pack into a crystal lattice. In order to do this they will need to be extended and not coiled, but the chance of a chain being fully extended is small. There will however be large portions of polymer chains that are extended. These will pack into an ordered crystal lattice and provide the microcrystalline regions while the tangled, coiled portions will form the disordered amorphous regions seen in Figure 4.9

In block copolymers (Sec. 4.3.8) it is possible to create a block that is highly crystalline together with one that is amorphous to obtain a final copolymer with special properties.

Crystalline polymers tend to have greater mechanical strength and higher melting point than amorphous polymers. Because the chains in the crystalline regions are closely packed they would also be expected to have higher densities, and this too is observed. For example, low density polyethylene has a tensile strength (Sec. 4.5.4) of 2000–2500 psi, a softening point of 85–87°C, and a specific gravity of 0.91–0.93. High density polyethylene, on the other hand, has a tensile strength of 3500–5500 psi, a softening point of 127°C, and a density of 0.94–0.97.

The degree of crystallinity of a polymer is measured by X-ray diffraction by the same technique used for single crystals. High density polyethylene may have as much as 90% crystallinity; the low density material has only about 55%, which is still high considering the extent of chain branching in the polymer. It occurs because the lengths of polymer chain between the branches are capable of getting close enough to other chains for crystallization to take place.

Crystallinity is also related to orientation. When a rubber band is stretched heat is generated. If one slowly flexes a wide rubber band, an appreciable amount of heat is generated which can be felt if the band is touched to the lips. This results from the friction of one polymer molecule rubbing against another as the stretching action causes them to align themselves. Before it is stretched the rubber band is largely amorphous. The alignment on stretching is tantamount to crystallization, which also causes the translucent rubber band to become opaque. Orientation of the polymer has caused it to crystallize. The heat generated can therefore also be thought of as heat of crystallization. The degree of crystallinity that can be induced in an elastomer by stretching can reach 30% (Fig. 4.10).

The orientation of crystallization of polymer molecules by stretching or drawing is an important step in the processing of polymers for use as textile fibers. When nylon and polyesters are manufactured they have low crystallinity. The stretching or drawing of the fibers causes the polymer molecules to line up or crystallize to give the longitudinal strength required in fibers.

A difference between fibers and rubbers is that the latter have much greater "elastic memory" and do not stay stretched and crystalline. In

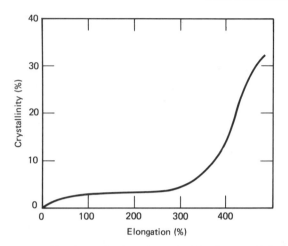

Figure 4.10 Crystallization of cured rubber by stretching.

other words, it is hard to stretch them beyond their limit of crystallinity. The reason is that rubber is lightly cross-linked, and these cross-links tend to pull the rubber molecules back to their original configuration. In addition, stretching is a disorder-order transition, and the entropy decreases. The return to the unstretched state is favored by the corresponding entropy increase. In nylon the "cross-linking" takes the form of hydrogen bonds that break and reform as the fiber is stretched. Indeed, stretching orients the molecule so that more hydrogen bonds are formed than are broken. Consequently the elastic memory is quickly exceeded on stretching.

4.5.2 GLASS TRANSITION TEMPERATURE, CRYSTALLINE MELTING POINT, AND SOFTENING TEMPERATURE

If we take an amorphous material such as polystyrene, melt it, and then allow it to cool it does not solidify sharply. First it goes from a viscous liquid to a rubbery solid, then to a leathery solid. Finally when all of the molecules have lost their thermal "wriggling" motions it becomes a glassy solid, recognizably polystyrene. The last change is a sharp one, and the temperature at which it occurs is called the glass transition temperature of the polymer T_g. Obviously it has a bearing on the properties of the polymer at the service temperature at which it is to be used. A polymer that is a soft, leathery material above its T_g may be a hard, brittle, amorphous one below the T_g.

The glass transition temperature is associated primarily with amorphous polymers, although crystalline polymers also have a glass transition temperature because all polymers have amorphous regions between the microcrystalline regions.

If a crystalline polymer is used above the T_g of the amorphous regions, the latter will be flexible, and the polymer will be tough. If the temperature is below the T_g, however, the amorphous regions will be glassy and the polymer brittle. A similar situation for block copolymers is described in Section 4.3.8.

The temperature at which a molten polymer changes from a viscous liquid to a microcrystalline solid is called the crystalline melting point T_m of the polymer. If the solid polymer is somewhat crystalline the change is accompanied by sudden changes in density, refractive index, heat capacity, transparency, and similar properties. It is analogous to the melting point of a nonpolymeric chemical compound but is not as sharp, and melting and freezing take place over a small range. The value of T_m depends on chain structure, intermolecular forces, and chain entanglement.

The softening point is an arbitrary measure of the temperature at which a polymer reaches a certain specified softness. It is of great importance as the upper service temperature of a polymer but has little significance on the molecular level.

Table 4.6 indicates typical values for the glass transition temperature

Table 4.6 Typical T_g's and T_m's of Polymers

	Temperature (°C)	
	T_g	T_m
cis-Polybutadiene	−101	4
cis-Polyisoprene	−73	29
trans-Polyisoprene	−58	70
Linear polyethylene	−70 to −20	132
Polypropylene	−16	170
trans-1,4-Polybutadiene	−9	139
Nylon 6,6	47	235
Poly(methyl methacrylate)	49	155
Poly(vinyl chloride)	70	140
Polystyrene	94	227
Polycarbonate	152	267
Cellulose triacetate	111	300
Poly(tetrafluoroethylene)	135	327

and crystalline melting point for various polymers. T_g is about one-half to two-thirds of T_m for most polymers if the temperatures are in degrees absolute. Deviations from this may be due to unusual molecular weight distributions, chain stiffness, and symmetry. These values, unlike the melting points of pure organic compounds, can be considered only "typical." The glass transition temperature may vary with the molecular weight of the polymer, its method of preparation, its end group distribution, and with the degree of crystallinity in a given polymer sample. For a completely unoriented material the glass transition temperature will be very low. When the material is oriented or converted to a crystalline state the glass transition temperature increases. The glass transition temperature for poly(ethylene terephthalate) may vary $-80–180°C$.

4.5.3 MOLECULAR COHESION

Molecular cohesion is the average force between the repeating units of a polymer chain and its neighbors. The forces are Van der Waal's forces or hydrogen bonds. Their magnitude can be calculated from cohesive energy density (Part II, Sec. 9.1).

Hydrogen bonds contribute most to molecular cohesion. Dipole-dipole forces contribute less, and London dispersion forces the least. The strength of the molecular interactions diminishes rapidly, actually with the sixth power of the distance between the molecules. Thus bulky amorphous polymers have relatively low intermolecular forces whereas those in crystalline polymers are much higher because the molecules are much closer together.

4.5.4 STRESS-STRAIN DIAGRAMS

Many of the quoted properties of polymers are derived from stress-strain diagrams. These are graphs of the deformation in a polymer sample (expressed as percent of elongation) produced by a particular applied stress (a tension expressed as pounds per square inch, grams per square centimeter, or meganewtons per square meter). Such diagrams can be generated quickly from a given polymer sample of controlled size in a testing laboratory. For reproducibility a standard sample and rate of extension must be used. An example (for a hard, tough plastic) is shown in Figure 4.11.

In the initial stages of the extension the stress-strain diagram is linear. That is to say the material obeys Hooke's law, and stress is proportional to strain. The gradient or slope of this straight section is called the initial modulus of elasticity and is a measure of the stiffness of the material (Young's modulus). If the applied stress is removed at this stage the polymer will return to its initial length. After a certain stress the graph

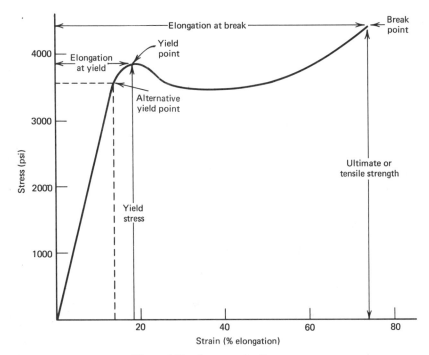

Figure 4.11 Stress-strain diagram.

ceases to be linear, and the extension is nonreversible; that is, a permanent deformation is produced. The yield point is defined as the maximum in the stress-strain curve, as shown in the diagram, and has an elongation at yield and a yield stress associated with it. Some authorities define the yield point as the point at which deformation becomes irreversible, that is at which Hooke's law ceases to be obeyed. Because many polymers do not obey Hooke's law the modulus is frequently expressed as pounds per square inch of tensile strength at a given degree of elongation such as 2%. This is called the 2% modulus. Some elastomers are better described by a 100% or 300% modulus.

After the yield point the plastic stretches relatively easily and, after stiffening a little toward the end, it breaks, and the curve comes to an abrupt end. The break point has associated with it an elongation at break or upper limit of extensibility and an ultimate tensile strength. The area under the curve up to the break point is called the work-to-break. The tensile strength is the strength required to pull the polymer apart, and the work-to-break is the work required to do it and is a measure of the ability of the polymer to resist not only tension or "pulling apart" but also other stresses such as bending, compression, impact, and twisting.

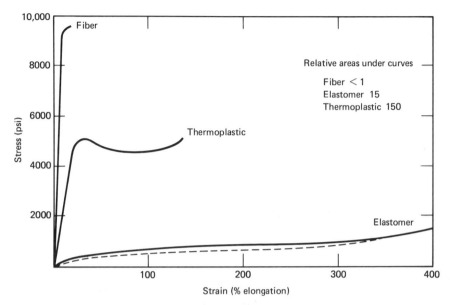

Figure 4.12 Stress-strain curves.

Figure 4.12 shows the type of stress-strain diagrams obtained for fibers, thermoplastics, and elastomers. Fibers, because they have been oriented, have high tensile strength and modulus and resist elongation. The work-to-break is small although the polymer is strong and resists "pulling apart." The elastomer, on the other hand, has high elongation but low tensile strength and modulus. Thus it has little resistance to deforming forces other than elongation. As its extension is reversible, a return stress-strain curve can be obtained if the stress is reduced before the sample breaks. This is indicated by the dotted line. The area within this hysteresis loop is the work dissipated as heat during the cycle.

The thermoplastic shown has quite a high modulus and tensile strength and a fairly high elongation. The area under the curve is large; hence this type of thermoplastic would be used if a wide variety of deforming forces were expected, whereas a fiber would have best resistance to simple tension. Elastomers are useful because they "bounce back" and can absorb energy by hysteresis.

4.6 CLASSES OF POLYMERS

Polymers are used in five main ways: as elastomers, plastics, fibers, surface coatings, and adhesives. Table 4.7 shows the combination of polymer properties required for the first three applications in terms of the four major properties.

Elastomers have a low modulus. Modulus is a measure of stiffness, and rubbers are not stiff. On the other hand, they need to be highly extensible, and elongations of 1000% are common. The crystallinity of an elastomer is low when the material is unstressed, but stretching leads to a higher degree of crystallization (Sec. 4.5.1). Molecular cohesion in elastomers must be low because otherwise the polymer chains will not easily slip over one another when the elastomer is stretched or slide back reversibly when the tension is released.

The second class of polymers is plastics. These are defined as polymers or resins that have been made into shapes, usually under pressure. Shaping may be done by processes such as molding, casting, extrusion, calendering, laminating, foaming, blowing, and many others. Nonetheless, the term plastic has a specific meaning in this context and should not be used to refer to polymers in general. The moduli, as well as other properties, of plastics vary widely according to their applications. For example, molding, calendering, or extrusion of a thin film or sheet requires a material with a low modulus so that the sheet is flexible. On the other hand, manufacture of a bleach bottle demands a stiff polymer—the stiffer the better so long as it is not brittle—so that the walls can be made thinner and material saved.

The extensibility demanded in elastomers is not needed in plastics,

Table 4.7 Polymer Properties

	Elastomers	Plastics	Fibers 150,000–
Modulus, (psi)	15–150	1500–200,000	1,500,000
Extensibility, upper limit (%)	100–1000	20–100	$<10^a$
Crystallization tendency	Low when unstressed	Moderate to high	Very high
Molecular cohesion (cal/ monomer unit)	1–2000	2–5000	5–10,000
Examples	Natural rubber Polychloroprene Polybutadiene etc.	Polyethylene Polypropylene Polystyrene Polyvinyl chloride Polyvinyl acetate	Silk Cellulose Nylon Dacron etc.

[a] Cellulose fibers have greater extensibility.

although a degree of extensibility is important so that the work-to-break is high enough for the plastic to resist twisting and impact. Generally the properties of plastics are intermediate between those of elastomers and fibers.

Fibers, if they are going to be knitted or woven into dimensionally stable garments, require unrelenting properties. They should have high modulus and low extensibility. The fibers must be strong so that a single thread will not "pull apart," and that requires a high modulus and molecular cohesion, properties closely related to high tensile strength and crystallinity.

It is more difficult to generalize about surface coatings and adhesives than about the other groups. A coating may require high extensibility and low modulus if it is to be applied to a soft rubber surface. On the other hand, a coating for a baked phenolic sheet may require low extensibility and high modulus. The property that both coatings and adhesives require is high adhesion, and the problem is to achieve this while maintaining a reasonable level of the diametrically opposed property of cohesion. In addition, a coating will require resistance to abrasion.

In general, coatings and adhesives tend to have low moduli, somewhere between those of elastomers and plastics. They must have some extensibility, particularly if the material is to be used on a dimensionally unstable substrate such as wood. It follows that coatings and adhesives have low crystallinity.

REFERENCES AND NOTES

P. T. Flory, *Principles of Polymer Chemistry*, Cornell University Press, Ithaca, NY, 1953 has hardly been superseded for many of its discussions of basic polymer chemistry. A standard work is F. W. Billmeyer, Jr., *Textbook of Polymer Science*, 2nd ed., Wiley, New York, 1971. Sound on polymer science and well referenced, it is nonetheless somewhat difficult to use. Another standard textbook, more technologically oriented, is D. C. Miles and J. H. Briston, *Polymer Technology*, Temple Press, London, 1965. An excellent text has been provided by B. Vollmert, *Polymer Chemistry*, Springer-Verlag, New York, 1973. Also useful are F. Rodriguez, *Principles of Polymer Systems*, McGraw-Hill, New York, 1970, and G. Odian, *Principles of Polymerization*, McGraw-Hill, New York, 1970. Very useful is a volume by W. R. Sorenson and T. W. Campbell, *Preparative Methods of Polymer Chemistry*, 2nd ed., Wiley-Interscience, New York, 1968. Preparation of numerous polymers is described. It provides a good basis for a laboratory course in polymer chemistry.

A celebration of Ziegler catalyst science and technology is contained in a book entitled J. C. Chien, Ed., *Coordination Polymerization, A Memorial to Karl Ziegler*, Proc. ACS Symp., UCLA, Los Angeles, CA, April 1974, Academic, New York, 1975.

Recent polymer research is summarized annually in H. J. Canton, Ed., *Advances in Polymer Science*, Springer, Berlin, 1958–1974. Fourteen volumes are now available. Selected newer polymers are reviewed in R. D. Deanin, Jr., Ed., *New Industrial Polymers*, ACS Symp. Series, No. 4, ACS, Washington, DC, 1974.

Sec. 4.2 Functionality has been described by Flory, *op. cit.*, p. 31 ff.

Sec. 4.3.3 Reactivity ratios of monomer pairs is described in detail in G. E. Ham, Ed., *Copolymerization*, No. 18 in High Polymers Series, John Wiley, New York, 1964.

Sec. 4.3.4 An excellent djscussion of molecular weight determination is contained in Flory, *op. cit.*, p. 266 ff.

Sec. 4.3.6 The May 1978 issue of the *J. of Phys. Chem.* is devoted entirely to articles on radical ions. The issue is in honor of Prof. M. Szwarc who did the early work in this field. P. Plesch (private communication) has recently proved that the first step in cationic polymerization by aluminum chloride is the disproportionation:

$$2AlCl_3 \rightleftharpoons AlCl_2^+ + AlCl_4^-.$$

Sec. 4.3.8 For a discussion of block polymers see D. C. Allport and H. Janes, *Block Copolymers*, Wiley-Interscience, New York, 1973.

Sec. 4.4.2 MDI (p,p'-methylene diphenylisocyanate) may also be prepared by a new process described in *Chemical Week*, November 9 (1977) 57 and *CHEMTECH*, **8** (1978) 382. Ethanol, nitrobenzene, and carbon monoxide are converted by a selenium catalyst to ethyl phenylcarbamate. This in turn is condensed with formaldehyde to give p,p'-methylene diphenyl diethyl carbamate. On pyrolysis, MDI and its oligomers are obtained and the ethanol is recycled. Apparently, this process is also applicable to tolylene diisocyanate.

Chapter Five
INDUSTRIAL CATALYSIS

The advent of cheap olefinic feedstocks in the early 1950s, together with all sorts of by-products, side streams, and intermediates, provided an exciting challenge for the industrial organic chemist. Many of the organic chemicals and polymers we have discussed in Part I and discuss in Part II were already being made before World War II. Few were being made from the same feedstock, and even after the change of feedstocks the methods of synthesis of the chemicals and polymers have changed.

The switch from traditional organic chemistry to modern "hot tube" industrial organic chemistry is the theme of Chapter 2, and we remarked frequently that the secret lay in the "appropriate" catalyst while glossing over its exact nature. We had little choice, for the mode of action of many industrial catalysts is obscure and the nature and exact formulation of most of them a closely guarded industrial secret. It is because catalysis is not fully understood and is something between black art and a recognizable branch of chemistry that it holds out such promise for the future. Certainly the recent successes of industrial organic chemistry have hinged on the development or discovery of new catalyst systems that have been more selective or more active than older ones or that have catalyzed new reaction pathways. It was estimated in 1966 that 70% of industrial processes involve catalysis, and the proportion has certainly increased since then. Before we speculate about the future in our final chapter, therefore, we provide some of the background to catalyst technology and the direction in which it is moving.

5.1 WHAT IS A CATALYST?

In many textbooks catalysts are ascribed four properties:

1 They change the reaction velocity but are not themselves altered and can be recovered unchanged at the end of the reaction.
2 They do not alter the position of equilibrium of a reaction.
3 They increase the forward and backward rates of a reaction proportionately.
4 They act by providing an alternate reaction pathway of lower activation energy.

Few industrial catalysts show this behavior precisely. Thus platinum catalyzes the combination of hydrogen and oxygen, but in the absence of a catalyst the reactants can apparently remain for many hundreds of years without reaction. It is meaningless to say that the platinum has merely increased the rate of reaction. It has brought it about.

Many catalysts cannot be recovered unchanged at the end of a reaction. When sulfuric acid catalyzes the esterification of ethanol with acetic acid the reaction products contain such materials as ethyl hydrogen sulfate, and the sulfuric acid has reacted with the by-product water formed in the reaction. This dilution side reaction is highly exothermic, and recovery of the sulfuric acid requires expenditure of chemical energy and reversal of the hydration and sulfate forming reactions. Because it alters the overall chemistry of the reaction the sulfuric acid catalyst alters the position of equilibrium and fails to increase the backward and forward rates proportionately.

The initiators of polymerization reactions are often called catalysts. They should not be, according to the above definition. They often appear as end groups in the final polymer and are not recoverable. Also, they do not affect the rate of reaction; they cause the reaction.

Heterogeneous catalysts may undergo sintering, etching, change of surface area, and poisoning. These may result from chemical reactions. The solid catalyst that does not have to be removed from the reactor every few months for regeneration is the exception rather than the rule.

In the manufacture of SNG by the catalytic rich gas process (Sec. 2.13) for example, the catalyst is nickel on alumina promoted with potassium. The alumina must be the high surface area γ-alumina, which has a cubic close-packed structure. When not in use as a catalyst, γ-alumina is stable up to 1100°C. In plant use, however, at temperatures as low as 400°C, the γ-alumina undergoes an irreversible phase change to a hexagonal close-packed structure, α-alumina or corundum. The collapse of the fine pore structure of γ-alumina and subsequent formation of corundum also leads to agglomeration of the nickel crystallites and loss of metal surface area so there is a drastic reduction in catalyst activity. The lifetime of the catalyst is governed by the rate of sintering of alumina and nickel.

Neither is it true that catalysts invariably act by lowering the activation energy. The rate constant k of a reaction can be represented by the Arrhenius equation $Ae^{-E/RT}$ where A is the preexponential factor, E the activation energy, R the gas constant, and T the absolute temperature. There are rare catalysts that act by lowering E and leaving A unchanged. Many more lower E and at the same time lower A, whereas a few raise E and still act as catalysts because they raise A by more than enough to compensate for the rise in E. In the acid catalyzed hydrolysis of methyl acetate E remains more or less constant while A increases. For different catalysts of the same reaction it often happens that a graph of log A against E gives a straight line. This remarkable phenomenon is called the compensation effect. It has been demonstrated for variations in A of up to 10^{14}, and its origin is still not satisfactorily explained.

Finally most textbooks overlook the crucial point of selectivity. A material that does not react at a measurable rate in the absence of a catalyst may react to give quite different products in the presence of different catalysts and at different temperatures and pressures. Ethanol passed over copper can give either acetaldehyde or ethyl acetate depending on the conditions. Passage over alumina gives ethylene or ethyl ether.

$$C_2H_5OH \xrightarrow{\text{Cu}} CH_3CHO + H_2$$

$$2C_2H_5OH \xrightarrow{\text{Cu}} CH_3COOC_2H_5 + 2H_2$$

$$C_2H_5OH \xrightarrow{\text{Al}_2O_3} C_2H_4 + H_2O$$

$$2C_2H_5OH \xrightarrow{\text{Al}_2O_3} C_2H_5OC_2H_5 + H_2O$$

A catalyst often permits a reaction to take place under milder conditions than would otherwise be possible, but operation at a lower temperature might lead to a different equilibrium and a different major product. Thus it may be better to define a catalyst as a substance that either brings about or accelerates one or several reactions in such a way that a large quantity of reactants is converted for a small amount of catalyst added. In this sense we might even consider the initiator of a polymerization reaction to be a catalyst. Alternatively, we could amend our definition to say that the catalyst should not appear in the final product. We could then say that metal complex catalysts are true catalysts while ionic and free radical initiators are not. The problem is largely semantic, but the inadequacy of the traditional definition should be appreciated.

Catalysts are conventionally divided into homogeneous and heterogeneous depending on whether they act throughout a single phase or at a phase boundary. Even this distinction has little theoretical significance. A homogeneous acidic catalyst like sulfuric acid acts in the same way as a heterogeneous acidic catalyst like silica-alumina. In certain examples of the Friedel-Crafts reaction it is uncertain whether catalysis is homogeneous or heterogeneous.

Homogeneous catalysts can be added easily to a reactant system but may be difficult to remove from the products. The catalytic effect is roughly proportional to the amount added. In heterogeneous catalysis it is not the mass of the catalyst that counts but its surface area. Preparation and pretreatment of solid catalysts is an art, and the activity of a catalyst depends on its previous history. Catalysts achieve high surface areas because of microporosity. Surface areas of $1000 \, m^2 g^{-1}$ are not uncommon. Although diffusion of reactants to a surface and of products away from it may present problems (and catalyst surfaces are rarely reproducible), heterogeneous catalysts have the great advantage of being easily separated from reactant systems. They are therefore the most widely used in industry. Homogeneous catalysts, on the other hand, are readily reproducible, and consequently they are the preferred systems for academic study. They have led to most of the insights we have into catalytic reaction mechanisms, but there is an obvious mismatch between what is possible for the research chemist and what is desirable from the industrial point of view.

5.2 HISTORICAL DEVELOPMENT

The development of catalysts seems for the most part to have followed social and economic needs rather than the leadership of theory. The art of catalysis has always been far ahead of the science, and catalytic theory owes much to industrial use of catalysis. There is a valuable interaction between the two, and the sort of catalyst screened by an industrial chemist will owe much to existing catalyst theory.

The earliest industrial catalysts were the enzymes used to convert grape juice to wine and wine to vinegar.

Catalysts for inorganic chemicals production emerged in the late eighteenth, nineteenth, and early twentieth centuries. Dr. John Roebuck, a consulting chemist to the metal industry in Birmingham, England, discovered in 1746 that the oxidation of SO_2 to SO_3 was catalyzed by nitric oxide, and he replaced the fragile glass vessels in which sulfuric acid had previously been made by lead chambers. In 1831 Peregrine Phillips, a Bristol vinegar manufacturer, found that platinum catalyzed the same reaction, but commercialization of the contact process did not take place in Europe until 1901 or in the United States, on a large scale, until after World War I.

The first heterogeneous catalyst to be used commercially was in the Deacon process, patented in 1868 and 1870 (Sec. 2.4.1), in which hydrogen chloride from the Leblanc process was oxidized to chlorine by passage with air over bricks soaked in cupric chloride.

Ostwald demonstrated the air oxidation of ammonia to nitrogen oxides (and hence to nitric acid) over a platinum catalyst in 1905, and Haber synthesized ammonia from nitrogen and hydrogen over a promoted iron oxide catalyst, the first plant coming on stream in 1913.

Industrial catalysis of organic reactions can be said to date from 1902 when Normann hardened fats by hydrogenating them over a nickel catalyst, a process based on the pioneering work of Sabatier. Methanol was produced from CO and H_2 over a zinc oxide-chromium oxide catalyst in Germany in 1923, and shortly afterwards naphthalene was oxidized to phthalic anhydride over a platinum catalyst. Vanadium pentoxide was later found to be more effective.

Catalytic cracking, widely used in the United States, was introduced by Houdry in 1936 and involved acidic catalysts such as silica-alumina.

A range of polymerization initiators and catalysts (Secs. 4.3.1; 4.3.6) was also discovered between World Wars I and II so that by 1939 examples of the first six classes of catalysts listed in Table 5.1 were in general use. During World War II Roelen, in Germany, developed the oxo (hydroformylation) reaction (Secs. 2.5.6; 5.5) in which hydrogen and carbon monoxide reacted with olefins to give aldehydes. The catalyst was a soluble cobalt-carbon monoxide complex $[Co_2(CO)_8]$, and the reaction, significantly, was the first industrial catalytic process to use transition metal complexes. The many subsequent examples of coordination catalysis include the Wacker and Ziegler processes and the metathesis reactions described in Chapter 2.

The other important advance since 1939 was the development of dual function catalysts. The first "platforming" catalyst was commercialized in

Table 5.1 A Summary of Catalyic Reactions [a]

Class	Functions	Examples
Metals	Hydrogenation Dehydrogenation Hydrogenolysis (Oxidation)	Fe, Ni, Pd, Pt, Ag
Semiconductors	Oxidation Dehydrogenation Desulfurization (Hydrogenation)	NiO, ZnO, MnO_2, Cr_2O_3, Bi_2O_3/MoO_3, WS_2
Insulators	Dehydration (Hydration)	Al_2O_3, SiO_2, MgO
Acids	Polymerization Isomerization Cracking Alkylation (Hydrolysis) Esterification	H_3PO_4, H_2SO_4, BF_3, SiO_2/Al_2O_3
Bases	Polymerization (Esterification)	Na/NH_3
Free radical initiators	Polymerization	$(C_6H_5COO)_2$ [b]
Transition metal complexes	Hydroformylation Polymerization Oxidation Metathesis	$[Co_2(CO)_8]$ $TiCl_4/Al(C_2H_5)_3$ $CuCl_2/PtCl_2$ WO_3, WCl_6
Dual Function Catalysts	Isomerization plus hydrogenation/ dehydrogenation	Pt on SiO_2/Al_2O_3
Enzymes	Varied	Amylase, urease, proteinases
Phase transfer catalysis	Reactions between immiscible reactants	$C_6H_5CH_2\overset{+}{N}(C_2H_5)_3Cl^-$ $(n\text{-}C_4H_9)_4N^+HSO_4^-$

[a] Data are largely drawn from G. C. Bond, *Heterogeneous Catalysis: Principles and Applications*, Clarendon Press, Oxford, 1974, (see notes). Less important functions are given in brackets.

[b] According to the broader definition of catalysts in Section 5.1.

1949 based on work by Haensel and co-workers. It consisted of a mixture of an acidic isomerization catalyst (SiO_2/Al_2O_3) and a hydrogenation/dehydrogenation catalyst (Pt) and could, for example, convert methylcyclopentane to benzene (Sec. 2.2.2). The range of catalysts now available is summarized in Table 5.1

5.3 CATALYSIS BY ACIDS AND BASES

The mechanism of acid and base catalysis is well understood. A typical acid catalyzed esterification reaction proceeds by the following mechanism:

$$
\underset{\text{Reagent acid}}{R-\overset{\overset{\displaystyle O}{\|}}{C}-OH} + H_3O^+ \xrightarrow{\text{fast}} R-\overset{\overset{\displaystyle O}{\|}}{C}-\overset{+}{O}H_2 + H_2O
$$

$$
R-\overset{\overset{\displaystyle O}{\|}}{C}-\overset{+}{O}H_2 + \underset{\text{Alcohol}}{R'OH} \xrightarrow{\text{slow}} R-\overset{\overset{\displaystyle O^-}{|}}{\underset{\underset{\displaystyle OH_2}{|}}{C}}-\overset{H}{\underset{R'}{O^+}} \xrightarrow{\text{slow}} H_2O + R-\overset{\overset{\displaystyle O}{\|}}{C}-\overset{H}{\underset{R'}{O^+}}
$$

$$
\Big\updownarrow \text{fast}
$$

$$
H_3O^+ + \underset{\text{Ester}}{R-\overset{\overset{\displaystyle O}{\|}}{C}-OR'}
$$

Similarly the iodination of acetone is catalyzed by acids because the latter hasten the keto-enol equilibrium.

$$
CH_3-\overset{\overset{\displaystyle CH_3}{|}}{C}=O + H_3O^+ \rightleftharpoons CH_3-\overset{\overset{\displaystyle CH_3}{|}}{\underset{}{C^+}}-OH + H_2O \rightleftharpoons CH_2=\overset{\overset{\displaystyle CH_3}{|}}{C}-OH + H_3O^+
$$

$$
\Big\downarrow I_2 \text{ (fast)}
$$

$$
CH_2I-\overset{\overset{\displaystyle CH_3}{|}}{C}=O + HI \longleftarrow CH_2I-\overset{\overset{\displaystyle CH_3}{|}}{\underset{\underset{\displaystyle I}{|}}{C}}-OH
$$

Brønsted and Lewis acids also act as catalysts by proton donation. For this reason, BF_3 is an initiator for cationic polymerization (Sec. 4.3.6) and is a catalyst according to our definition in Sec. 5.1. The cracking catalyst, SiO_2/Al_2O_3, is a Brønsted acid because the Al^{3+} ion replaces the Si^{4+} ion in the SiO_2 lattice, leaving one unit of charge short, which is compensated for by a labile hydrogen ion. $HCl/AlCl_3$ catalyzes the Friedel-Crafts reaction between benzene and propylene (Sec. 2.5.3) by the following mechanism, which has been confirmed by isotopic labeling:

$$CH_3CH=CH_2 + AlCl_3 + HCl \rightleftharpoons [AlCl_4^-][CH_3 \overset{+}{C}HCH_3]$$

$$CH_3-CH=CH_2 + H^+ \rightleftharpoons CH_3-\overset{+}{C}H-CH_3$$

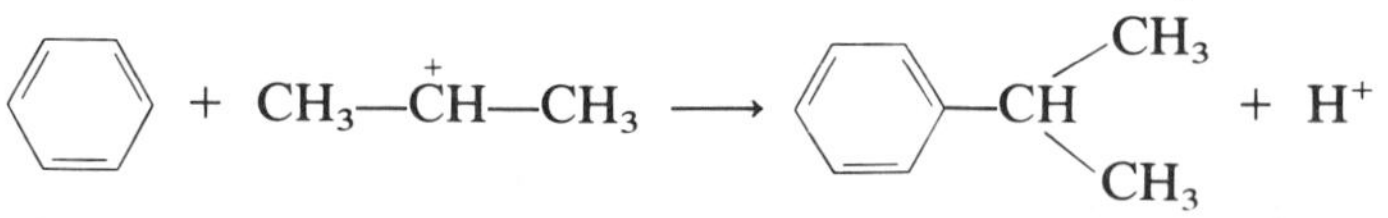

Catalytic cracking proceeds by ways of carbonium ions (Sec. 2.2.2).

Basic catalysis is rarer in industry than acidic catalysis. Examples include one-shot phenol-formaldehyde resins (Sec. 4.4.1) and isocyanate formation (Sec. 4.4.2). The opening of an epoxide ring frequently depends on a basic catalyst. An example is the reaction of ethylene oxide with acrylic acid to give hydroxyethyl acrylate, a trifunctional monomer used in baking enamels. The catalyst is a tertiary amine or quaternary ammonium salt.

$$CH_2=CHCOOH + CH_2CH_2 \longrightarrow CH_2=CHCOOCH_2CH_2OH$$
$$\underset{O}{\diagdown\diagup}$$

Whether a catalyst is acidic or basic can alter the reaction products. Alkylation of toluene with ethylene in the presence of a basic catalyst gives side chain alkylation, whereas acidic catalysis results in ring alkylation to give *o*- and *p*-ethyltoluene (Fig. 5.1).

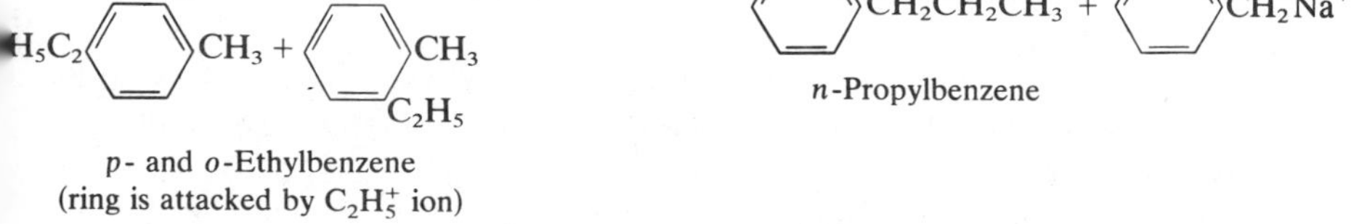

Figure 5.1 Alkylation of toluene.

Similarly methanol and carbon monoxide give acetic acid in the presence of an acid catalyst and methyl formate in the presence of a base (Sec. 2.4.3).

$$CH_3OH + CO \begin{array}{c} \overset{\text{acid}}{\nearrow} \quad CH_3COOH \\ \\ \underset{\text{base}}{\searrow} \quad CH_3\!-\!O\!-\!\underset{\underset{O}{\parallel}}{C}\!-\!H \end{array}$$

5.4 CATALYSIS BY METALS, SEMICONDUCTORS, AND INSULATORS

The majority of heterogeneous catalysts are metals and metal oxides. They may alternatively be classified as p- and n- type semiconductors and insulators. At the surfaces of these materials reactants can adsorb. Physical adsorption is weak ($\Delta H \simeq 40$ kJ mole^{-1}) and does not lead to catalytic activity. Chemisorption (dissociative adsorption) on the other hand is strong ($\Delta H \simeq 400$ kJ mole^{-1}) and the adsorbents themselves dissociate and form chemical bonds with the surface.

In the Haber process, for example, the reactants both adsorb and dissociate on iron. Hydrogen dissociates freely even at liquid air temperatures, but nitrogen does not do so until about 450°C, and this is the rate-determining step. Once the nitrogen molecules have dissociated (with adsorbed atoms written Cat≡N where Cat is the catalyst) the atoms can react readily with neighboring hydrogen atoms to give Cat=NH, Cat—NH$_2$, and finally Cat $\cdots$ NH$_3$ from which the NH$_3$ is easily desorbed.

Table 5.2 gives the heats of adsorption of nitrogen on various surfaces. On glass and aluminum only physical adsorption occurs, and these materials do not catalyze ammonia production. Iron, tungsten, and tantalum all give dissociative adsorption, but the preferred catalyst is iron because it has the smallest heat of adsorption, and therefore the products are most easily desorbed.

The foregoing simple theory of catalysis by way of chemisorption provides a reasonable explanation for catalysis by metals. The mode of action of "pure" metal oxides and nonstoichiometric metal oxides is more complicated, and any attempt at explanation had to await a quantum mechanical theory of solids and the application of this to heterogeneous

**Table 5.2   Heats of Adsorption of
Nitrogen on Various Surfaces**

Surface	ΔH (kJ mole^{-1})
Glass	~ -7
Aluminum	~ -42
Iron	-293
Tungsten	-397
Tantalum	-585

catalysts. Together with the crystal and molecular orbital ligand field theories these now provide some theoretical underpinning for the catalytic effects of semiconductive metal oxides.

In the broadest terms we can say that catalytic activity is related to the state of the d bands, corresponding to the assembly of d orbitals at the catalyst surface. The crystal surface of a semiconductor may be thought of as having a supply of electrons and a supply of "holes" where electrons can locate themselves. These either donate electrons to adsorbed molecules or draw them out. Thus they participate in reactions as free valences so that the addition of a heterogeneous catalyst to a reactant system is in some ways like the addition of free radicals. It is interesting to note that oxidation reactions are catalysed by p-type semiconductors that have surplus "holes," whereas hydrogenations are brought about by n-type semiconductors that have excess electrons. This fits in with a definition of oxidations as reactions in which electrons are lost and reductions as reactions in which they are gained. Insulators are effective at dehydration.

Another way of looking at this is to think of catalysts as weakening chemical bonds either by feeding electrons into antibonding orbitals on adsorbed molecules or by withdrawing them from bonding orbitals.

The theory of heterogeneous catalysis by semiconductors is complicated and is still far from satisfactory. The nature of adsorbed species is being examined by a range of new techniques including electron energy loss vibrational spectroscopy, X-ray photoelectron spectroscopy, ultraviolet photoelectron spectroscopy, Auger electron spectroscopy, and electron-stimulated desorption. Fundamental knowledge about the mechanism of heterogeneous catalysis may make it possible one day to tailor-make a catalyst that functions as efficiently and selectively as the semiconductor devices used in modern electronics. It may also be possible to minimize problems of catalyst poisoning either by modification of catalyst structure or by the admixture of antidotes to the feedstock or intermittently to the catalyst.

5.5 COORDINATION CATALYSIS

Coordination catalysis has provided many of the dramatic synthetic advances since World War II. The first industrial example was the oxo reaction in which an α-olefin was treated with CO and H_2 at 150°C and 200 atm. in the presence of a cobalt catalyst. The CO reacts with the cobalt catalyst to give the hydrocarbon-soluble complex $[Co_2(CO)_8]$, dicobalt octacarbonyl, which is believed to participate in the sequence shown in Figure 5.2. It is generally agreed that alkyl and acyl cobalt carbonyls are intermediates and that $H-Co(CO)_3$ or $H-Co(CO)_4$ are active species. Other aspects of the mechanism are more doubtful.

Other industrial examples of coordination catalysis include the range of Ziegler-Natta catalysts (Sec. 4.3.10), which lead to stereoregular products, and the asymmetric synthesis of levodopa (Part II, Sec. 8.13.2), which leads to an optically active product.

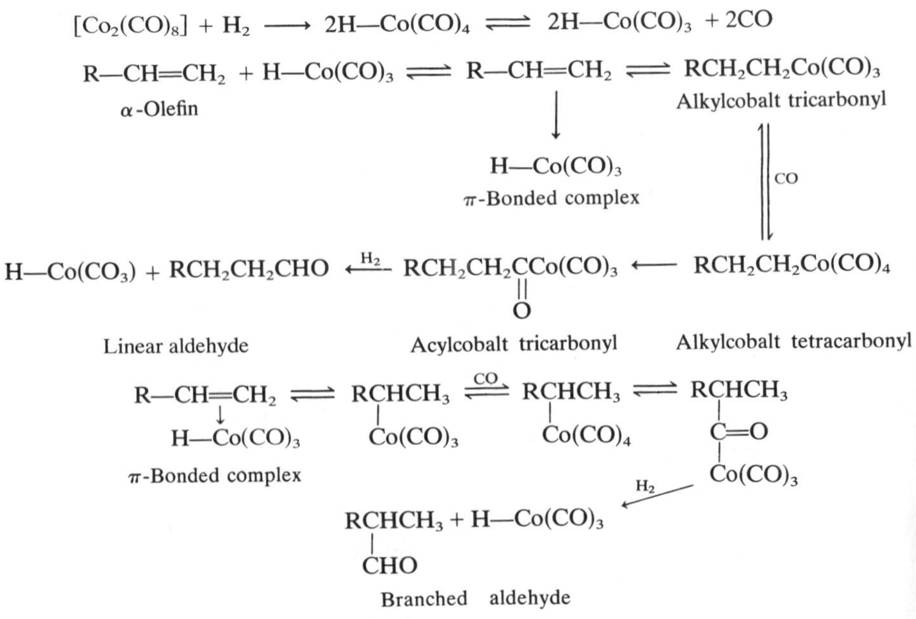

$$[Co_2(CO)_8] + H_2 \longrightarrow 2H-Co(CO)_4 \rightleftharpoons 2H-Co(CO)_3 + 2CO$$

$$R-CH{=}CH_2 + H-Co(CO)_3 \rightleftharpoons R-CH{=}CH_2 \rightleftharpoons RCH_2CH_2Co(CO)_3$$

α-Olefin

Alkylcobalt tricarbonyl

$$H-Co(CO)_3$$
π-Bonded complex

CO

$$H-Co(CO_3) + RCH_2CH_2CHO \xleftarrow{H_2} RCH_2CH_2CCo(CO)_3 \longleftarrow RCH_2CH_2Co(CO)_4$$
$$\underset{O}{\overset{||}{}}$$

Linear aldehyde   Acylcobalt tricarbonyl   Alkylcobalt tetracarbonyl

$$R-CH{=}CH_2 \rightleftharpoons RCHCH_3 \xrightarrow{CO} RCHCH_3 \rightleftharpoons RCHCH_3$$
$$H-Co(CO)_3 \qquad Co(CO)_3 \qquad Co(CO)_4 \qquad C{=}O$$

π-Bonded complex

$Co(CO)_3$

$$RCHCH_3 + H-Co(CO)_3 \xleftarrow{H_2}$$
$$CHO$$

Branched   aldehyde

Figure 5.2 Mechanism of oxo reaction.

The Wacker process (Sec. 2.4.3) proceeds by way of a coordination complex as do the various processes now being carried out over rhodium catalysts such as the very important oxo reaction that leads to linear products (Sec. 2.5.6), and the carbonylation of methanol to acetic acid (Sec. 2.4.3), which is believed to proceed via a methyl iodide intermediate and a rhodium/iodine/carbon monoxide complex:

$$CH_3OH \xrightarrow{\ HI\ } CH_3I + H_2O$$

A recent intriguing example is duPont's synthesis of aniline by the direct amination of benzene (Sec. 2.7.3).

A final example is the duPont synthesis of 1,4-hexadiene, a possible diene for use in ethylene-propylene terpolymer rubbers (Part II, Sec. 4.2.3). Ethylene and butadiene are passed into a solution of rhodium chloride in ethanolic hydrogen chloride or an acid solution of a nickel phosphite complex. The total mechanism is complicated, but the following steps have been definitely established. If the catalyst is written NiL_4, the L signifies the ligand, $P(OC_2H_5)_3$.

$$NiL_4 \underset{}{\overset{H^+}{\rightleftharpoons}} HNiL_4^+ \underset{+L}{\overset{-L}{\rightleftharpoons}} HNiL_3^+ \xrightarrow{C_4H_6} C_4H_7NiL_3^+ \underset{-L}{\overset{+L}{\rightleftharpoons}} C_4H_7NiL_2^+$$

$$\Big\Uparrow C_2H_4$$

$$NiL_2 + H^+ + CH_2{=}CHCH_2CH{=}CHCH_3 \longleftarrow C_4H_7NiL_2^+(C_2H_4)$$

1,3 Hexadiene

The butadiene-nickel-ligand-ethylene complex has the approximate structure

$$\left[\begin{array}{c} \begin{array}{c} CH_2 \\ \| \\ CH_2 \end{array} \longrightarrow \begin{array}{c} L \\ | \\ Ni \\ | \\ L \end{array} \longleftarrow \begin{array}{c} CH \\ \diagup \diagdown \\ CH \\ \diagdown \\ CH_2 \end{array} \begin{array}{c} CH_3 \end{array} \end{array}\right]^+$$

in which the two hydrocarbons are bonded to the nickel atom, as ligands, through their double bonds.

Coordination catalysis can take place either homogeneously or heterogeneously. In the metathesis of olefins, for example, WO_3 will act as a heterogeneous catalyst whereas WCl_6 plus $C_2H_5AlCl_2$ in ethanol will work in solution. Similarly in the Wacker process for vinyl acetate (Sec. 2.4.2) the mechanism was worked out for the homogeneous reaction, but the system proved too corrosive, and the workable industrial process finally involved a heterogeneous catalyst.

The advantage of all this is that catalyst mechanisms can be studied in solution under reproducible conditions, but the eventual catalyst can be used in any convenient form. This adds to the already impressive potential of coordination catalysts.

5.6 DUAL FUNCTION CATALYSIS

Dual function catalysts are mixtures of two catalysts each of which performs differently. The most widely used dual function catalysts are the mixtures of an acidic isomerization catalyst (SiO_2/Al_2O_3) with a hydrogenation/dehydrogenation catalyst (Pt). They are used for catalytic reforming ("platforming") described in Section 2.2.1 in which paraffinic and naphthenic feedstocks are converted to aromatics. Typical feedstocks are cyclohexane and methylcyclopentane. A dehydrogenation catalyst can convert cyclohexane to cyclohexene and then to benzene, and an acidic catalyst can convert methylcyclopentene to cyclohexene. Only a dual function catalyst can convert methylcyclopentane to benzene. For this transformation the two types of catalytic sites are required with transfer of an olefinic intermediate between the sites. The same effect could not in many cases be achieved by successive beds of the two catalysts. In a reaction of the type, $A \rightleftharpoons B \rightarrow C$, in which the equilibrium of the first reaction lies to the left, two beds of single function catalyst would not bring about reaction because not much of B would be formed in the first bed. On a dual function catalyst, however, the small quantity of B is removed as soon as it is formed. More A consequently changes to B, and the reaction to give C is accomplished.

The distances between the different catalytic sites govern the effectiveness of a dual function catalyst. n-Heptane will reform over a mixture of catalysts of particle size 1.0–10 μ but not 100–1000 μ.

5.7 ENZYMES

Enzymes are the oldest industrial catalysts. In certain respects they are also among the newest. Enzymes are biological catalysts, and many show absolute specificity. For example urease will only catalyze urea hydrolysis [$OC(NH_2)_2 + H_2O \rightarrow CO_2 + 2NH_3$]. Some enzymes will attack certain chemical groups wherever they occur (group specificity). For example, proteolytic enzymes will split the peptide linkage. Proteolytic enzymes will only attack peptides made up either from L-amino acids or from D-amino acids, and this is called stereochemical specificity.

Enzymes are proteins but may be associated with nonproteins (coenzymes or prosthetic groups) essential to their activity. Activity is usually related to a small region of the molecule referred to as the active center.

At low concentrations of substrate the rate of enzyme action is directly proportional to both enzyme and substrate concentration. If the concentration of substrate is raised, however, the rate ceases to increase and becomes independent of substrate concentration. Thus enzymes are only efficient in dilute solution. Furthermore, they operate only under a limited range of pH (rarely <4) and temperatures (usually $<50°C$), and reactions become very slow as $0°C$ is approached. Nonetheless, molecule for molecule, enzymes are much more effective than nonbiological catalysts. They appear to act by forming a complex with a molecule of reactant. The latter is bound to two active sites on the enzyme molecule, and one of these "pushes" electrons while the other "pulls" them. This concerted action is believed to be the basis for enzyme catalytic efficiency.

Fermentation is of course based on enzymes (Sec. 3.4). Enzymes are used in isolated applications such as the addition of proteolytic enzymes to detergents (Part II, Sec. 7.6.8). Sewage treatment depends in part on enzymes. Immobilized enzymes are used in the production of fructose (Sec. 3.4.1) and penicillin (Part II, Sec. 8.7.3). Although fermentation was an important route to chemicals between World Wars I and II, enzymes have not recently found application in large tonnage chemical production.

5.8 CATALYSTS OF THE FUTURE

There can be few uncatalyzed reactions that remain to be discovered. Thus consideration of future organic chemical processes is in the end a discussion of catalysts of the future. Neither can we envisage many new applications for acid and base catalysis, which has been known and understood for a long time except perhaps in the area of molecular sieves. The fields where new catalysts might emerge are semiconductor catalysis, coordination catalysis, enzyme catalysis, dual function catalysis, and just possibly some entirely new area.

Advances in semiconductor catalysis are likely to involve improvements in efficiency and selectivity. Most reactions that offer hope for the future will depend to some extent on present catalysts, and the aim is to get more of a particular product. This is not as pedestrian as it sounds. In 1976 Mobil improved their p-xylene catalyst to give a 10% increase in capacity. Overcapacity already existed because of the slowdown in the growth of polyester fiber, and the price of p-xylene decreased even more than it would otherwise have done. In 1978 it dropped below the o-xylene price for the first time ever, which was not what Mobil had intended but which was in small part due to their increase in catalyst efficiency.

Similarly, the first ammoxidation catalysts for acrylonitrile (Sec. 2.5.2) gave 6% yields—clearly not commercial. Bismuth phosphomolybdate raised this to 65%, and a plant was built in 1959. A uranium oxide catalyst gave yields of 80% in 1966. A subsequent proprietary catalyst has raised this even further. The new catalyst is also more active, and plant throughput has been raised by perhaps 20%. A new Union Carbide process for low density polyethylene uses a catalyst that operates at relatively low pressures and saves 70% of the energy consumed in the traditional process.

Dual function catalysis started impressively with the "platforming" reaction, but there have been few developments since then. Oxychlorination involves two processes, but the chlorination reaction is uncatalyzed. Ammoxidation similarly involves two reactions (oxidation of propylene to acrolein followed by further oxidation to acrylic acid). In the Distillers' Company's process these were carried out separately with acrolein as an intermediate whereas in the SOHIO process they were carried out simultaneously. There was no evidence, however, for dual function catalysis with intermediates moving from site-to-site on the catalyst surface. There are also several processes in which a number of reactions occur on the same catalyst; for example, the nickel catalyst in duPont's HMDA synthesis (Sec. 2.6.2) causes the double bond to shift and then catalyzes the addition of the HCN across it. This is not dual function catalysis, however, on the above definition. While we expect to see more processes in which several steps take place in a single reactor, or on a single catalyst, the development of more true dual function catalysts does not seem as likely as it once did.

Neither are we optimistic about the future of enzyme catalysis. Fermentation will always have its application, and when oil supplies run out the world may be driven back to fermentation ethanol feedstocks for aliphatic organic chemicals. The snags are, first, there is slight possibility in the medium term future of our being able to synthesize enzymes, and we are therefore stuck with those supplied by nature; and second, the sensitivity of enzymes to concentration and temperatures means that reactions are slow, require large tanks, and the products are inevitably in dilute solution. In the case of beer or wine this is not important, and distillation costs are trivial compared with value added in whisky and brandy. In the harsh world of organic chemicals, however, the cost of recovering usable products from dilute solution in an energy-expensive world is a drawback that will only be overcome by a dramatic breakthrough in enzyme technology. Immobilized enzymes offer great prospects for the future if their technology can be made practical, and there is much research in progress in the United Kingdom on continuous fermentation processes of which the production of single cell protein from methanol was the first example (Sec. 3.4).

Novel catalysts may emerge from further developments in molecular sieves. The dehydration of appropriate alkali metal aluminosilicates (known as synthetic zeolites) forms a crystalline structure with intracrystalline cavities, and the access to these is by small pores of uniform size. This pore size can be controlled to some extent by variation in the chemical composition and method of processing of the molecular sieve. Molecular sieves are used for separation processes, the desired compound forming a clathrate compound with the sieve (e.g., the Molex process, Part II, Sec. 7.6.7; the Parex process for separation of p-xylene from mixed xylenes, Sec. 2.9).

Molecular sieves may also have catalytic acid sites, however, and their introduction in 1964 transformed catalytic cracking (Sec. 2.2.1). They give much better gasoline yields than the traditional silica/alumina catalysts, being more active, more selective, and more resistant to poisoning. As their design becomes better understood, it becomes possible to make more sophisticated catalysts and a new zeolite ZSM5 illustrates the potential power of such catalysts. Mobil found ZSM5, a high acidity, high Si:Al zeolite, in 1968 but failed to recognize its activity for 5 years. They then found it would convert methanol by way of dimethyl ether to gasoline plus mixed aromatics up to durene (1,2,4,5-tetramethylbenzene) (see note to Sec. 2.12.2). It also provides aromatics from paraffins and oxygenated material and can thus be used to increase to octane number of gasoline. Fischer-Tropsch hydrocarbons (Sec. 3.1.2) have an octane number of about 60 and passage over ZSM5 can raise this to 90. Mobil is piloting a fluidized bed unit for use by Sasol in South Africa in conjunction with their Fischer-Tropsch plant. This may represent the cheapest route to gasoline from coal.

ZSM5 also improves the efficiency of xylenes isomerization and catalyses the disproportionation of toluene to benzene and xylene (Sec. 2.8).

Phase transfer catalysis is finding increasing industrial application. It is worthy of mention not because it is likely to be involved in the manufacture of large tonnage heavy organic chemicals but because it is an unusual and elegant catalytic technique that is economical with energy and gives high yields at low residence times and under mild conditions. It is therefore typical of the methods that will be sought in the future.

It finds application where reactants are immiscible. An example is the production of penicillin esters (Part II, Sec. 8.7.3) in which the free carboxyl group of the aminopenicillanic acid is esterified with a labile group that will hydrolyze in the gut. The desired reaction is:

AMP—COO⁻K⁺ + [structure] → [structure] + KCl

Potassium salt
of ampicillin   Talampicillin

but the ampicillin salt is water-soluble whereas the acid chloride dissolves in organic solvents and would be hydrolyzed by water. Mild conditions are essential to avoid decomposition of the acid chloride and the β-lactam ring of the ampicillin. The acid chloride is therefore dissolved in an organic solvent such as methylene chloride or chloroform and brought into contact with an aqueous solution of the ampicillin salt. A phase transfer catalyst such as tetrabutylammonium chloride is added to the aqueous phase.

The tetrabutylammonium cation is lipophilic and migrates into the organic layer as an ion pair carrying with it the ampicillin anion, which is the less hydrophilic of the anions present. Esterification of the ampicillin anion takes place smoothly at 25°C in the organic layer, and the tetrabutylammonium ion pairs with the chloride ion that is generated. The latter is so hydrophilic that it carries the tetrabutylammonium cation back into the aqueous phase and the procedure is repeated until esterification is complete.

Phase transfer catalysis is also used industrially for synthesis of polycarbonates (Part II, Sec. 2.1.8) oestradiolcarbamate and a range of fine chemicals and is likely to be of value in the future in similar areas.

Although many of the above areas of catalyst development are of great interest, coordination catalysis is undoubtedly the glamor area at present and in the immediate future. We have already devoted considerable space to it. In many cases coordination catalysts operate under conditions as mild as enzymes without the latter's drawbacks. The field appears to be versatile and to offer a nice balance of theory and empiricism. While technologists are not having to screen random compounds unguided by theory, there are still sufficient aspects that are not understood to provide scope for imagination and enterprise.

We cannot ignore the possibility of something entirely new and un-expected, but it is idle to make predictions about what is by definition unpredictable. There are, however, a few types of reactions that could possibly be widely applied if suitable technology could be developed.

One such group of reactions is oxidation or reduction by oxygen or hydrogen carriers that operate at moderate temperatures and pressures. An example is the ethylanthraquinone route to hydrogen peroxide, which is carried out at 40°C and 1–3 atm:

Ethylanthraquinone   Ethylhydroanthraquinone

A catalytic reaction that generates hydrogen from water at relatively low expenditure of energy is already at commercialization stage. The crucial step is a nickel catalyzed decomposition of ammonium iodide

$$2NH_4I + Ni \xrightarrow{400°C} NiI_2 + 2NH_3 + H_2$$

$$NiI_2 \xrightarrow{800°C} Ni + I_2$$

$$I_2 + Na_2CO_3 \xrightarrow{6-700°C} 2NaI + CO_2 + \tfrac{1}{2}O_2$$

$$2NaI + CO_2 + 2NH_3 + H_2O \longrightarrow Na_2CO_3 + 2NH_4I$$

Total $H_2O \longrightarrow H_2 + \tfrac{1}{2}O_2$

Overall thermal efficiency is expected to exceed 30%.

REFERENCES AND NOTES

Books on catalysis suffer from one of two faults—either they propose numbers of complicated theories scarcely supported by experimental evidence, or they list hundreds of reactions and the catalysts used without much attempt to explain catalyst action. An excellent general account is given in *Kirk-Othmer*, and the chemical engineering aspect is summarized in R. H. Perry and C. H. Chilton, *The Chemical Engineer's Handbook*, 3rd ed., McGraw-Hill, New York, 1973, Sec. 4, p. 29.

There are two authoritative multivolume series: P. H. Emmet, Ed., *Catalysis*, Reinhold, New York, 1954–; and W. G. Frankenburg, E. K. Rideal, V. I. Komarewsky, P. W. Selwood, and P. B. Weisz, Eds., *Advances in Catalysis*, Academic Press, New York, 1948–. Less well established is H. Heinemann and J. Carberry, Eds., *Catalysis Reviews*, Dekker, New York, 1963–.

Standard books on heterogeneous catalysis include G. C. Bond, *Catalysis by Metals*, Academic Press, New York, 1962; J. M. Thomas and W. J. Thomas, *Heterogeneous Catalysis*, Academic Press, London, 1967; C. N. Satterfield and T. K. Sherwood, *The Role of Diffusion in Catalysis*, Addison-Wesley, Massachusetts, 1963; and more recent, R. Avis, *The Mathematical Theory of Diffusion and Reaction in Permeable Catalysts*, 2 vols., Clarendon Press, Oxford, 1975. The last three books are fairly mathematical in their treatment. A shorter book with an academic slant is A. J. B. Robertson, *Catalysis of Gas Reactions by Metals*, Logos, London, 1970, while G. C. Bond describes some of the standard catalytic reactions important in industry in a paperback, *Heterogeneous Catalysis: Principles and Applications*, Clarendon Press, Oxford, 1974. Table 5.1 is largely drawn from this source.

Homogeneous catalysis is tackled in the relatively new review series R. Ugo, Ed., *Aspects of Homogeneous Catalysis*, Carlo Manfredi, Milan, 1970–. Both types of catalysis are covered in B. Delmon and G. Jannes, Eds., *Catalysis—Heterogeneous and Homogeneous*, Elsevier, 1975.

Three books with a practical slant are *Catalyst Handbook*, ICI Agricultural Division, Springer Verlag, Berlin, 1970; C. L. Thomas, *Catalytic Processes and Proven Catalysts*, Academic Press, New York, 1970; and O. V. Krylov, *Catalysis by Non-Metals*, Academic Press, New York, 1970, which contains rules for catalyst selection.

Sec. 5.1   For a not too technical introduction to the subject see V. Haensel and R. L. Burwell, Jr., "Catalysis," *Scientific American*, **225,** No. 6, (1971), 46. The figure for the proportion of industrial reactions depending on catalysis came from G. M. Dixon and F. E. Shepherd, *Catalysis—Art or Science*, Gas Council Research

Committee 136, London, 1966. The spread of "A" values in the compensation effect is reported by F. S. Feates, P. S. Harris, and B. G. Reuben, *JCS Faraday Trans.*, **70** (1974) 2011. The irreversible degradation of CRG catalysts is documented in A. Williams, G. A. Butler and J. Hammonds, *J. Cat.*, **24** (1972), 352; W. H. Gitzen, *Alumina as a Ceramic Material*, p. 35, Special Publication #4, American Ceramic Soc. Columbus, OH (1970); and G. Yamaguchi and H. Yanagida *Bull. Chem. Soc. Japan*, **36** (1963), 1155. It was brought to our attention by C. Komodromos of British Gas Corp.

Sec. 5.3   The classic book on acid-base catalysis is R. P. Bell, *The Proton in Chemistry*, Cornell University Press, 1959. The neglected area of basic catalysis has recently been covered in H. Pines and W. M. Stalick, *Base Catalyzed Reactions of Hydrocarbons and Related Compounds*, Wiley, New York, 1977. New physical techniques for investigation of the properties of adsorbed layers are summarized by J. T. Yates, *Chem. Eng. News*, August 26, 1974, p. 9.

Sec. 5.5   The duPont hexadiene synthesis is described by C. A. Tolman, *J. Am. Chem. Soc.*, **92**, (1977) 6777. See also "New Perspectives in Surface Chemistry and Catalysis," *Chem. Soc. Rev.*, **6**, (1977) 373. The mechanism for the carbonylation of methanol appeared in D. Forster, *J. Am. Chem. Soc.*, **98** (1976), 846.

Sec. 5.7   Enzyme catalysis seems even stranger to the traditional chemist than the wave-mechanics-of-solids theories of heterogeneous catalysis. A sound introduction is provided by M. L. Bender and L. J. Brubacher, *Catalysis and Enzyme Action*, McGraw-Hill, New York, 1973. More advanced is K. Tamaru and M. Ichikawa, *Catalysis by Electron Donor-Acceptor Complexes—Their General Behavior and Biological Roles*, Halsted Press, New York, 1975.

Sec. 5.8   "Horizons in Catalysis" by J. D. Idol, *Chem. & Ind.* (1979), p. 272 is an excellent general article on the development of molecular sieve catalysis.

There has been a number of recent articles on academic aspects of phase transfer catalysis—for example, G. W. Gokel and W. P. Weber, *J. Chem. Ed.*, **55** (1978), 350 and 429; J. M. McIntosh, *J. Chem. Ed.*, **55** (1978), 235. Industrial aspects are covered by B. G. Reuben and K. Sjöberg, *CHEMTECH*, to be published.

The nickel-catalyzed ammonium iodide decomposition was reported in *Chem. Econ. Eng. Rev.*, **9**, (1977) 25 and commented on in *CHEMTECH*, **7**, (1977) 741.

Chapter Six
THE FUTURE OF THE CHEMICAL INDUSTRY

The period between the end of World War II and 1973 saw a steady and unprecedented rise in the standard of living in developed countries. It was also a time of rapid expansion and technological change in the chemical industry. The events of 1973 and the world recession that followed them were seen by many as a sign that the party was over and that the dream of sustained economic growth would be destroyed, if not by depletion of oil supplies then by increase of pollution, the population explosion, or a deficiency of water or of food supplies. Others more optimistic saw the recession as a temporary setback on the road to an economic millennium in which cheap, inexhaustible nuclear energy would replace fossil fuels by the time the latter ran out. Which scenario is correct is still not clear, and in the final chapter of this book we propose to review briefly the causes for the growth of the chemical industry between 1945 and 1973 and to speculate about the medium-term future, the next 20 to 30 years. In this period oil will still be available, albeit at greatly increased prices, and the breakthrough to power by way of nuclear fusion will not yet have occurred.

Figure 6.1 shows the expansion of manufacturing as a whole and of the chemical industry in the United States, United Kingdom, and Japan. The

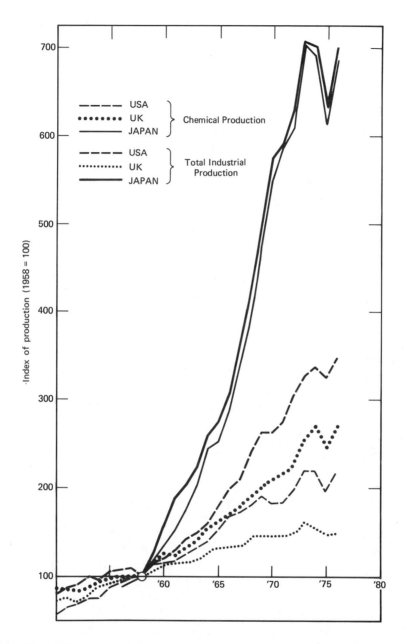

Figure 6.1 Expansion of manufacturing and of the chemical industry. Indices of production for the United States, United Kingdom, and Japan, 1950–1976.

takeoff point for the chemical industry in the United States was during World War II. In the United Kingdom it was 1951, when the first steam-naphtha crackers were commissioned, and in Japan it was about 1960.

The chemical industry in the United States and W. Europe has expanded at about twice the rate of the manufacturing industry, but that growth has not been across the board. Architectural coatings, for example, must be applied to something that has either been built or manufactured. Thus the architectural coatings industry is tied to the construction industry and has expanded at a similar rate. Lubricating oils only partly shared in the automotive boom because improved oils used in engines with closer tolerances gave better performance and needed to be changed less frequently. Most heavy inorganic chemicals were being made on a large scale before 1939, and there were no dramatic new uses or breakthroughs in technology to cause their market to expand. The sectors that performed spectacularly were petrochemicals (heavy organic chemicals, plastics, elastomers, fibers) and pharmaceuticals.

6.1   THE PETROCHEMICAL REVOLUTION

Prior to 1950 the majority of aliphatic organic chemicals in the United Kingdom were derived from ethanol made by fermentation of molasses. Even in the United States, which was relatively advanced, the 85% of aliphatics obtained by fermentation in 1925 had only dropped to 28% by 1945. Aromatic organic chemicals were obtained from coal tar. There were three prerequisites for the petrochemical revolution—an understanding of the nature of high polymers, a source of cheap olefins and aromatics, and the development of chemical engineering skills necessary for a large scale chemical industry.

The science of high polymers was developed between World Wars I and II by Hermann Staudinger, R. H. Kienle, and W. H. Carothers. Before then polymers were regarded as aggregates of molecules held together by vaguely defined secondary valence forces. By 1930 the functionality rules had been defined, and X-ray crystallography had shown the partly crystalline nature of high polymers. By 1939 the majority of the commercially important, high volume plastics of today were being manufactured in the United States, Britain, and Germany. Ethylene when required was made by dehydration of fermentation ethanol. Acetylene, derived from limestone and coke by way of calcium carbide, was an important raw material, particularly for vinyl chloride and vinyl acetate.

The earliest thermoplastics were fabricated on machinery developed for use with rubber. Fabrication methods specific to thermoplastics such as blow molding came later, and it is possible that, without the rubber processing equipment, the commercialization of thermoplastics would have been delayed by many years.

World War II gave tremendous impetus to the organic chemicals and polymer industries. The US synthetic rubber program has already been mentioned (Sec. 2.4.4; Part II, Sec. 4.1), and Germany too produced huge quantities of synthetic rubber for her war machine. Polyethylene was developed as an electrical insulator. PVC was used for raincoats, curtains, and shoes. Nylon was brought to the production stage. Although the new plastics had many defects and the Allies laughed at the "ersatz" materials to which the Germans had been reduced, the manufacture of synthetic polymers emerged from the war as an established and important industry.

6.1.1 THE SWITCH TO PETROLEUM AND NATURAL GAS

The fact remained that synthetic polymers, especially the more easily processed thermoplastics, were relatively expensive because they were made from high cost starting materials like acetylene and ethylene. Another source of olefins already existed in the United States. The US petroleum industry had long operated cracking processes to convert the higher boiling petroleum fractions into gasoline, and it was known that large quantities of C_2—C_4 olefins could be produced. Propylene was already available from refinery off-gases. The first petrochemical plant had been commissioned in 1923. Propylene was treated with sulfuric acid and the sulfate ester hydrolyzed to isopropanol. The US chemical industry moved toward a natural gas base in the 1930s. The 3 million lb of ethylene produced in 1930 had increased to 26 million by 1940, most of it being converted to ethylene glycol. The amounts of refinery off-gases available also increased because of growth of the automobile industry and consequent demand for gasoline.

The situation in Europe was quite different. European manufacturers in the 1930s were bound by governmental policies of national self-sufficiency. Lacking indigenous petroleum they took no interest in petrochemicals until after the war, and even then they faced a different situation (see notes). Natural gas had not yet been discovered, and cars were few compared with the United States. From the oil refiners the market demanded fuel oil for heating and a relatively small amount of gasoline. The intermediate-boiling naphtha fraction was too volatile for the kerosene market and insufficiently volatile for gasoline. It was virtually unsaleable and provided a cheap, readily available feedstock for petrochemicals that the oil men were practically prepared to give away.

Thus the European and subsequently the Japanese chemical industries relied on naphtha for their petrochemicals whereas the United States relied on the small proportion of ethane and propane in natural gas. This situation has continued until the present day, although, as described in Chapter 2, the transatlantic industries seem to be drawing closer together.

The availability of cheap olefins removed the last obstacle to the petrochemical revolution. Synthetic polymers were used on an increasing scale, as were other petrochemical-based products such as synthetic detergents. The early prejudice against plastics as an inferior substitute for the natural God-given article was not entirely unfounded. The early polyethylene bowls stress-cracked and would not withstand boiling water; PVC baby pants embrittled after a few weeks as the plasticizer was extracted by urine and soapy water. But "Formica" laminates were obviously better than wood for working surfaces in kitchens, and unbreakable long-playing PVC phonograph records were clearly superior to the old, fragile 78 rpm discs. Silk and semisynthetic rayon could not compete with man-made nylon stockings, and although synthetic fiber garments were in general less comfortable than natural fibers their easy care properties won them rapid acceptance in a society where women sought to decrease time spent on household chores.

Synthetic polymers grew because the standard of living was rising, because they replaced traditional materials, and because they created their own markets. Microcircuits for example were never "potted" in anything but synthetic polymers. Polymer growth was also helped by the low cost of petroleum, improvements in chemical engineering, and the development of new processes, factors discussed in the following paragraphs. Some details of the growth are shown in Figure 6.2. On a volume basis polymers are now used on the same scale as iron and steel. The growth of petrochemical feedstocks and the way in which feedstocks for organic chemicals have changed are shown in this figure.

6.1.2 OIL EXPLORATION

During the 1950s and 1960s the oil companies arrived at an understanding of the geological principles underlying oil formation, and thus a high proportion of test drillings produced positive results. The companies had no incentive to cap the wells and keep the price of petroleum high. Rather they were eager to exploit the wells, sell the oil, and recover their investment in exploration as quickly as possible. The price of Middle East crude in real terms dropped steadily between 1950 and 1970, and in spite of huge increases in world oil consumption the known world recoverable reserves of oil increased annually over the same period. As a consequence the real price of petrochemicals fell.

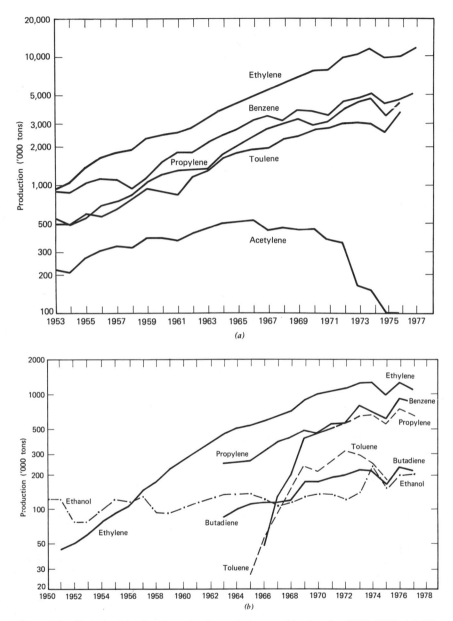

Figure 6.2 Growth of feedstock production and change of feedstocks, 1950–1977. (*a*) US production of some feedstocks, 1953–1977. (*b*) UK production of some feedstocks, 1951–1977. Figures for ethylene and propylene are for material for use in chemical manufacture. Benzene and toluene figures exclude coal-based material. Benzene production includes material made by dealkylation of toluene, and toluene figures include the material dealkylated.

245

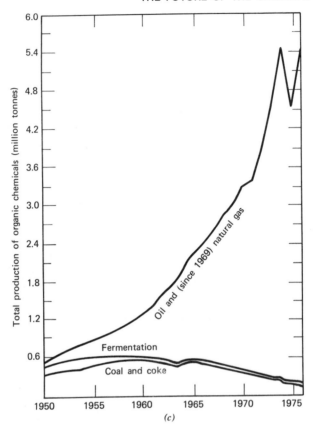

Figure 6.2 (c) Sources of production of UK organic chemicals.

Furthermore, before World War II, oil was refined in the producing country, and the finished products shipped overseas. It was often possible for the refiner to supply the various markets of the world, which then differed widely, in such a way as to avoid waste.

In postwar Europe, however, there was a trend toward refining petroleum in consumer rather than producer countries, and this too made the by-product hydrocarbon feedstocks for cracking available at low prices.

6.1.3 THE CHEMICAL ENGINEERING REVOLUTION

Another factor that helped to reduce the price of petrochemicals and synthetic polymers was the remarkable economy of scale that proved possible in the chemical industry. Chemical engineering came of age after World War II, and its most striking achievement was the change from batch to continuous operation in most chemical plants. This led to economies of scale. The size of plants escalated. Figure 6.3 shows the capacity of UK naphtha crackers since 1950. The growth in capacity of a single cracker from 70 million lb of ethylene year^{-1} in 1952 to 1 billion lb year^{-1} in 1968 is a mark of the extent to which chemical engineers responded to the petrochemical challenge. No single steam cracker with a capacity much greater than this had been built by 1978, but Shell has announced a cracker for 1981 with a capacity of 1.5 billion lb ethylene year^{-1}. The leveling off in the rate of increase in plant capacity suggests that no dramatic economies of scale in ethylene crackers are at present accessible.

The increase in plant size has been based the square-cube law (Sec. 1.3.4) and on changes in equipment. For example, bubble cap distillation columns have been largely replaced by sieve tray columns, which are cheaper and give higher throughput for lower pressure drop. Sieve trays had been known since the nineteenth century, but only in the last 25 years have design data been good enough to make them the dominant pattern.

Analytical methods for the continuous monitoring required by continuous plants have been devised. Gas chromatography, cheap, robust mass spectrometers, and a host of spectrometric methods together with advances in electronics have made possible the computerized plant of today. The transition away from the nineteenth century chemical plant where thermometers, gauges, and valves were scattered throughout the plant and an important valve had its own operator was already in progress by World War II. Simple feedback systems were already in use. Some plants had control rooms with pipes, gauges, and important valves inside it, and remote control was beginning to appear. Nonetheless, it was a far cry from the present systems where plants with elaborate feedback mechanisms are run by a computer and are supervised by a few technicians in an air conditioned control room. It is for this reason that labor costs in the chemical industry are small.

These engineering changes in polymer production were matched by economies of scale for the production of heavy organic chemicals needed for monomers and also at the level of polymer fabrication. The first thermoplastics were fabricated by methods devised for the rubber industry. The growth of thermoplastics went hand-in-hand with new methods for their fabrication. Such techniques as blow molding of bottles, injection molding, and the extrusion of thin film increased production speeds of plastic articles, reduced costs, and led to larger markets.

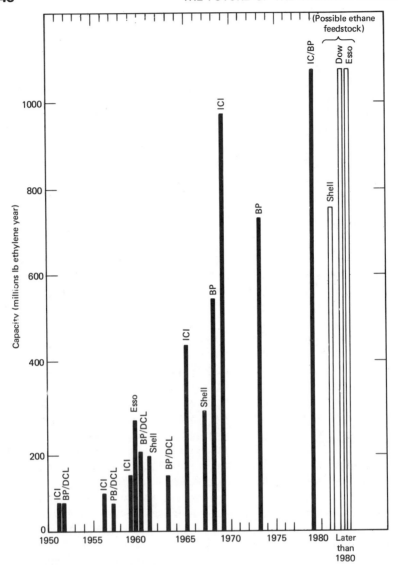

Figure 6.3   Capacity of UK naphtha crackers since 1950.

6.1.4   THE 1973 CRISIS

Between 1960 and 1972 the price of ethylene, which was already a cheap, large tonnage chemical, dropped from 5.0 to 3.0 cents lb^{-1} because of a low inflation rate, improved processing, and the economies of scale inherent in larger and larger production units. Corresponding propylene

Table 6.1 **Prices of Ethylene and Propylene**

Year	Ethylene (cents lb^{-1})	Propylene (cents lb^{-1})
1953	5.0	—
1956	5.0	1.0
1959	5.0	1.8
1962	4.7	2.0
1965	4.0	2.1
1968	3.4	2.3
1971	3.0	2.7
1973	3.3	2.8
1976	11.75	8.75
April 1979	13.9	9.75

prices varied between 1.8 and 2.7 cents lb^{-1} during this period (Table 6.1).

By the early 1970s a large gap had opened up between the marginal cost of oil production and the "value" of the oil, that is the cost of acceptable substitutes. Oversupply had pushed the price down near the marginal cost. The OPEC cartel restricted supply and enforced sharp rises in petroleum prices after 1970 and most spectacularly in 1973. By 1976 inflation, shortages, and the fivefold increase in oil prices had boosted the price of ethylene to 11.75 cents lb^{-1} and of propylene to 8.75 cents lb^{-1}. The petrochemical industry's growth ceased in 1974, but by 1977 growth had started again and capacity was being expanded. There was evidence of dissension within the cartel, and there were signs that in the short term the real price of petroleum was dropping again. This situation was abruptly reversed in the aftermath of the Iranian revolution and prices moved sharply upwards in June 1979. In the longer term they can only increase further, and this is discussed in Section 6.3.

6.2 SOCIAL AND ECONOMIC PRESSURES ON THE CHEMICAL INDUSTRY

The chemical industry depends for its success not only on the technical ingenuity of chemists and chemical engineers but also on the social and economic context in which it operates. Processes of elegance and apparent usefulness may be superseded or discarded for reasons that could not have been foreseen. For example, the process for steam reforming of naphtha to give a feedstock for ammonia production was discovered in the United Kingdom. It operated successfully for a few years and then became obsolete because the discovery of natural gas under the North Sea made available a cheaper, sulfur-free feedstock.

The interaction of chemistry and society goes much deeper than this. Supplies of raw materials are affected by political factors. The products that the chemical industry makes are determined by market factors, and its ability to make them is governed by regulations on toxicity, health, and pollution. Furthermore, the ability of the chemical industry to expand or even to replace worn-out plants depends on its cash flow and the level of profits that competition, government regulations, and public opinion permit it to make. The chemical industries of countries in the Western World and Japan must compete for finance with other ways in which the same money could be spent. Should a government raise taxes to finance higher defense or welfare expenditures, or should it allow companies to keep more of their incomes in the hope that they will reinvest the money in a way that will create new jobs for the unemployed?

The free world, as we write, is in disarray and is weakening politically while its members try to export their unemployment and inflation to each other. The $40 billion yearly surplus that the OPEC countries are accumulating is perturbing the international political and monetary systems that have existed since World War II. A drain on financial resources of this magnitude may mean a long term recession in the free world quite different from the slump of 1975 which, as we have seen, was due largely to hoarding. Voices are to be heard asking whether the institutions of the free world can cope with the new situation and whether democracy itself is not bound to go under because it encourages rising expectations that can no longer be satisfied.

Even in the centrally planned economies, however, decisions about allocation of resources have to be made. They are made not by the complex interaction of producers, consumers, voters, and government agencies but by a collection of bureaucrats who are able to decide not only what is produced but also what is bought. There is no evidence that the latter system can produce prosperity or an environment of the quality enjoyed by free societies, or that it can harness the talents of its population to achieve the same rate of technological advance. On the contrary, 60 years after the Bolshevik revolution in Russia it is clear that the economic and political systems it spawned are not only repressive but also inefficient. The "butter mountains" or the "wine lakes" created by overenthusiastic farmers of the European Economic Community may testify to the imperfections of the Western world's economic system, but it is a success story in comparsion with the persistent failure of Russian agriculture to provide adequate grain harvests to feed its population.

Only in the field of armaments can the centrally planned economies claim equality.

The chemical industry is not isolated from political upheaval. If the West were to adopt another economic system through internal revolution, either by the extreme right or the extreme left, the pattern of the chemical industry would change. If there were nuclear war (a scenario fashionable in the fifties but less fashionable now) then the chemical industry set up by the survivors would be different. If the Russians or Chinese were to conquer the West or the Arabs to buy it, the structure of our society would change markedly. For a forecast to have any meaning we must assume a moderately stable political system. Given that, there seem to be four major areas where the chemical industry will face problems in the next quarter century. They are energy and availability of raw materials, toxicity, pollution and safety, and overcapacity and shortage of capital. We discuss them in turn.

6.3 ENERGY AND THE AVAILABILITY OF RAW MATERIALS

Modern industrial society is energy intensive. Instead of walking we travel in vehicles. Instead of the vehicle being pulled by a horse that eats grass, a resource replaced with the aid of solar energy, we power it by burning nonrenewable resources of gasoline. In our whole way of life we use the power of machines to replace the puny efforts of human muscles. The average US industrial worker has the energy of 244 men at his disposal. But the energy that powers the machines comes largely from combustion of fossil fuels laid down over millennia. Since they are finite they will at some point be used up. That is the fundamental energy crisis.

There are other, shorter term, energy crises. The major long term reserves of crude oil are located in the Middle East, an area notorious for its political instability. Possession of the oil gives the rulers of these sparsely populated desert areas a degree of political leverage whose magnitude is uncertain because it has not yet been tested. It also means that a large amount of the cash flow of the West and Japan (and of the developing countries) is flowing into the bank accounts of these rulers with implications for the world monetary system that, again, are still not clear.

The long term question is how soon available reserves of fossil fuels will be used up, and no confident answer can be given because no one knows how large the undiscovered reserves are. Every year until 1970 known world oil reserves increased in spite of spiraling levels of consumption. They have since started to drop. Visionaries give precise dates as to when they will run out, but these depend on dubious estimates of undiscovered reserves and even more dubious extrapolations of energy consumption in the future. Let us say merely that within the lifetimes of ourselves or our children crude oil will become a rare and expensive commodity. Natural gas will run out sooner, but coal reserves will last for several hundred years.

Hydroelectric power is renewable, but the scope for hydroelectric schemes based on present technology is limited unless energy prices go much higher. Nuclear fission power might tide us over for 50 years, but world supplies of uranium are finite, and it requires more energy to extract uranium from many low grade ores than is recovered by "burning" it in a nuclear pile. Interest seems to have waned in the fast breeder reactor, which produced one nuclear fuel as it burned another. A more serious problem is that the early reactors are coming to the end of their working lives and we are faced with the problem of what to do with wastes that will be dangerous for the next 25,000 years, a period two and one-half times that which has elapsed since neolithic man and the dawn of civilization.

Nuclear fusion—the process that keeps the sun hot—has not yet been demonstrated on earth even on a pilot plant scale, and while it would provide an almost infinite supply of cheap nonpolluting energy, there is no guarantee that the formidable technical problems will ever be overcome.

Many other sources of renewable energy have been suggested—solar power, wind power, the energy forest (forests of fast growing trees surrounding a wood-fired power station), geothermal power, tidal power, and others. Liquid hydrogen, a much canvassed possibility, is only an energy carrier, and to produce it would require either cheap nuclear power, as discussed, or a catalyst system that would enable sunlight to photolyze water. Reserves of oil locked up in tar sands and oil shale are larger than liquid oil reserves, but their exploitation is fraught with problems. None of these sources has the flexibility and convenience of oil and natural gas, and energy from them will cost more perhaps by a factor of 2–5 times. SNG might even be an order of magnitude more expensive. This does not mean civilization will come to an end or that we will necessarily go back to living in caves, but it will mean a high level of substitution and a change in the social attitudes that were developed in an energy rich world. We shall have to work hard even to maintain our present standard of living. Inhabitants of developing countries in a way will suffer less in that they live in less energy intensive societies; on the other hand, the hopes of many of them of reaching an economic "take-off" point are likely to be disappointed.

The United States and the chemical industry are both large users of energy. The 5% of the world population who live in the United States consume one-third of the world's energy. If manufacturing industries are arranged in order of energy intensiveness (see notes) then all chemical and allied products industries except toiletries, pharmaceuticals, and "plastics products not elsewhere counted" come in the top quartile. If we include the chemical process industries (e.g., cement, the most energy intensive of all, and glass) in our definition of the chemical industry, then chemicals consume one-third of the energy of all manufacturing industry.

We have already pointed out that the chemical industry makes good use of its energy. Value added is high. Bottles made from plastics, for example, are less energy intensive than glass bottles, although oil is used for them both as raw material and a source of energy. Thus the chemical industry is in a good position to compete for any supplies of oil that exist.

We give below an estimate of the effect on European costs of chemicals, plastics, and manufactured products of a 100% increase in the price of naphtha. It appears that the increase is "diluted" about ten times by the time the product reaches the consumer.

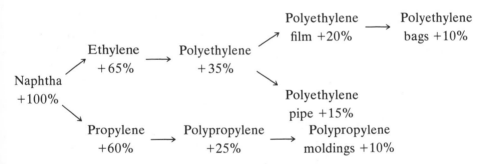

The chemical industry can view these figures with a degree of equanimity. On the other hand, its dependence on energy makes it vulnerable. Falling standards of living would lead to reductions in demand for chemicals, which in turn would reduce economies of scale and increase the price of chemicals even more. This would be the opposite of the benificent spiral that brought plastics to their present position. Packaging is a large volume outlet for plastics, and it is easy to envisage a scenario where paper and cellophane recovered some of their lost markets and the retailing industry became more labor intensive and less packaging intensive.

In the medium term, however, even though rising oil prices will hinder growth in the chemical industry, they are unlikely to cause it to alter dramatically. It is significant that the fivefold increase in the price of crude oil between 1970 and 1974 slowed growth and increased costs and prices but did not lead to an abrupt shift in technology. The only examples we can find of processes that may be rendered uneconomical by the fivefold increase are the electrothermal route to phosphorus and phosphoric acid and the Aman route to periclase (see notes). World capacity may be expected to drop as power costs rise, as present plants need replacement, and as long term fixed price contracts for cheap power expire, but even that is not certain (see notes). Dearer oil may alter the balance between natural and synthetic rubber and make synthetic "natural" rubber (Part II, Sec. 4.1) less economical.

In process technology the trend is likely to be away from the high temperature, high pressure engineering projects of the interwar years. Mild conditions will be the aim, and Ziegler-Natta catalysis the model. Separation processes currently account for a high proportion of the chemical industry's energy demands, and we may expect to see the increased interest in processes like reverse osmosis and ion pair extraction, which demand little energy, and away from distillation, which consumes a lot. A major advantage of the new route to acetic acid from methanol and carbon monoxide (Sec. 2.4.3) is that the yield is about 99.9% based on methanol, and the product requires little purification.

For the organic chemicals industry the question of availability of raw materials is inseparable from that of energy. All chemical feedstocks can alternatively be burned as fuel. This applies not only to the much discussed petroleum and coal but also to such products as bagasse from sugar cane, which is conventionally burned to provide energy for sugar refining but which is also an excellent source of cellulose for fiber board and similar products.

Over the next generation while petroleum is still available it would seem sensible for the chemical industry to use oil as a feedstock while coal is used for energy. We have already suggested that market forces will see to it that this will happen if political forces do not override them. In the longer term the chemical industry will reluctantly have to make do with coal, oil shale, tar sands, and vegetable products. Its ability to compete with food growers for available land is likely to be less than its ability to compete with energy users for oil, and it is frightening to envisage a situation where pressure on land for food might be so great that the chemical industry will be unable to produce agrochemicals to keep food yields high.

In the final analysis, if abundant energy is available, the organic chemicals industry can use almost any feedstock. For example. carbon dioxide from calcination of limestone will react with hydrogen from electrolysis of water to give carbon monoxide, and carbon monoxide-hydrogen mixtures will react to give hydrocarbons of the kind found in petroleum. It is the energy that is the problem.

6.4 TOXICITY, POLLUTION, AND SAFETY

In general, mankind has not shown much concern for its environment. This is scarely surprising. The occasions when environmental dangers have been recognized by people powerful enough to do something about them have been few compared with times when fear of conquest or pressure of starvation overrode longer term considerations. Joseph succeeded in convincing Pharoah of the dangers of overcultivation and made him make provision for the seven lean years that would follow the seven fat ones; but he was the exception.

Ignorance and malice also played their part in the assault on the environment. It may have been Helen of Troy's face that "launched a thousand ships and burned the topless towers of Ilium" but it was the trees of Greece that were chopped down to make the ships, and the much-admired civilization of classical Greece that deforested that section of the Mediterranean basin and allowed the erosion of its soil. At the eastern end of the Mediterranean the "fertile crescent" of biblical times was deforested by the Ayyubids (to prevent the Crusaders from being tempted back), by the Turks (as a punishment to their subjects for not paying taxes), and by the Mamlukes (who disapproved of all trees except olive trees). The soil structure in turn was destroyed by nomadic farming, especially the raising of goats. The bedouin, romantically known as sons of the desert, could more properly be described as fathers of the desert.

Similarly in the United States poor agricultural methods led to the creation of the dust bowl, and there is more than an even chance that the current destruction of forests in Brazil to create more agricultural land will, fairly rapidly, lead to the encroachment of desert. This century has seen a few attempts to reclaim land, and the efforts of the Tennessee Valley Authority, the Israelis, and the Dutch are worthy of mention.

Industry, even primitive industry, brought with it other kinds of environmental damage. The effluents from butchers, chandlers, tanners, and the like polluted water courses. It was no longer true as it once had been that a river purified itself every few miles.

With the coming of the industrial revolution, the problems were compounded. The burgeoning glass, soap, and paper industries required alkali, and this was supplied by way of the Leblanc Process. Salt was heated with sulfuric acid to give hydrogen chloride and sodium sulfate (salt cake). The salt cake was heated with coal and limestone to give calcium sulfide slag (black ash) and sodium carbonate, which was extracted and purified. The hydrogen chloride was discharged from tall chimneys, and the calcium sulfide slag dumped in heaps where it reacted with the dilute hydrochloric acid produced whenever it rained. The discharge of HCl into the atmosphere or hydrochloric acid into water courses was forbidden in the United Kingdom by the Alkali Act of 1863, an early milestone in pollution control. There were many who condemned this interference with business, but the clamor died down when increasing demand for bleaching powder made the HCl a saleable by-product.

In addition to industrial effluents there were problems with adulteration of food, either deliberate or accidental. In a celebrated case in England before World War I, a number of children died from arsenic poisoning from eating candy containing glucose. It turned out that the glucose was made by hydrolysis of starch with sulfuric acid and that the manufacturer had used acid made from arsenic-containing pyrites by the lead chamber process instead of contact process acid that was arsenic-free.

There were also widespread industrial diseases and injuries. Girls making matches from yellow phosphorus developed necrosis of the jaw ("phossy jaw") and textile workers developed "mule spinners" cancer across the lower abdomen and thighs from contact with carcinogenic lubricating oils as they leaned over their machines. As late as World War I women working in munitions factories developed severe jaundice from inhaling dust from the trinitrotoluene used as an explosive.

These examples illustrate the three kinds of environmental problems arising out of the chemical industry. There is the genuine accident, the product that is inherently dangerous, and the manufacturing process that menaces people who work in the plant or live near it. Many more historical examples could be given to illustrate the point that environmental problems have existed for millennia and were not newly discovered in the mid-1960s.

Nonetheless, the last twenty years have seen a change of attitude of people in the developed countries toward industry in general and the chemical industry in particular. In the United States this dates possibly from the passage of the Delaney amendment in 1958 and the "great cranberry crisis" of 1959. This arose because of misuse of the cranberry weed killer, aminotriazole. It is of low toxicity, but in the long term can lead to tumors of the thyroid. Residues were found on marketed cranberries, which were consequently seized by the FDA, and millions of Americans were deprived of cranberry sauce with their Thanksgiving dinners.

While this incident and others like it are widely publicized, the benefits brought about by the chemical industry are rarely appreciated. We comment in Part II, Section 1.13 that there is a gap between producers and consumers of chemicals that results in widespread ignorance of the role of chemistry in a modern society. Chemists are seen by many as demonic white-coated figures who devise new ways of making the planet uninhabitable. Haven't they given us thalidomide, nerve gases, DDT, defoliants, saccharin, and nonbiodegradable plastics? Even penicillin contributes to the problem of overpopulation.

There are at least three reasons for this change of mood. First, the population of developed countries in the mid-1960s became unprecedentedly wealthy. A high proportion of them for the first time in history had the leisure to bother about environmental considerations instead of where the next meal was coming from. Second, the advances in analytical chemistry made it possible to estimate trace quantities of contaminants that would previously have gone undetected (sensitivities have increased by 10^4 to 10^6 over 20 years). And third, the improvements in communications and analysis of data brought about by the computer made it possible to pick up hitherto unsuspected correlations.

Which environmental problems are genuine problems and which are trivial and based on misunderstandings of the nature of science? Alas, there is no acid test. Everything is a matter of judgment, but correct judgment is not as easy as is sometimes pretended. There is no reason to suppose that the world would be a better place without the chemical industry, but on the other hand this does not give the industry a right to wreck the environment. Environmentalists should remember that a chemical plant worker has about 15 times as much chance of being killed while driving to work as he has of dying at work. Equally, management should not use this as an excuse for failure to try to eliminate what risk there is.

We shall refer in Part II to a number of environmental problems—vinyl chloride and angiosarcoma of the liver (Part II, Sec. 6.2.1), detergents and the pollution of water streams (Part II, Sec. 7.6.7), and the thalidomide tragedy (Part II, Sec. 8.13). To show the scope of the problems the chemical industry faces we here list a selection of other disasters that have damaged its image.

6.4.1 FLIXBOROUGH

At 4:53 p.m. on Saturday, June 1, 1974, the Flixborough (UK) works of Nypro (UK) Ltd., a joint company of the National Coal Board and Dutch State Mines, was demolished by an explosion of warlike dimensions. Because it was not a working day the death toll was smaller than it would otherwise have been. Twenty-eight were killed and 36 injured. Outside the plant there were 53 injured; 1821 houses and 167 factories and shops were also damaged. The plant was engaged in the production of nylon 6 from cyclohexane that was catalytically oxidized to a cyclohexanol/one mixture (Sec. 2.7.2). One of the reactors in the oxidation train had been removed for repair of a leak, and the gap had been bridged with a 20-inch diameter dogleg pipe that was inadequately supported. The key post of plant engineer was vacant, and there was no proper design study; in fact, no one realized that one was necessary. There was creep cavitation and zinc embrittlement of the stainless steel of which the tube was made, and a crack developed due to nitrate stress corrosion. This corrosion was created because nitrate-treated cooling water had been used in the past to dilute small leakages of cyclohexane from the plant. Through the leak poured cyclohexane starting at least at 8.6 bar and 155°C. It formed a massive cloud that ignited, destroying the plant and much of the surrounding neighborhood. The source of the ignition was not established, but it was thought to have been in the hydrogen plant. The committee of inquiry commented on the coincidence of unlikely errors and cleared from blame the fabricators and installers of the bypass line and the control staff. They noted the absence of a plant engineer and regretted that the narrowness of engineering education had prevented any other engineer on site from realizing potential dangers.

6.4.2 MINAMATA BAY

In 1953 and 1965 a strange disease was reported in the area around Minamata Bay in Japan. Its symptoms were lack of coordination, narrowing of the field of vision, and sometimes death. Animals and birds were affected as well as humans. Forty-six people died, and 120 became seriously ill. It was established that the symptoms were those of mercury poisoning and that the victims had consumed contaminated seafood. The mercury came from a plastics factory upstream from the bay and was concentrated by sea creatures in their systems. Mercury itself is not very poisonous, but apparently microorganisms present in the bottom muds of aquatic environments can convert mercury and its salts to the dangerous dimethyl mercury, which then becomes concentrated by way of the food chain.

In a similar incident at Niigata, Japan, the fish were found to contain 5–20 ppm of mercury. Japan now has a limit of 0.01 ppm dimethyl mercury in discharge waters.

6.4.3 SEVESO

On July 10, 1976, there was an explosion at the Icmesa plant at Seveso, near Milan, in northern Italy. Six and one-half hours after the end of a run there was an unexpected rise in temperature in a reaction vessel. This caused an increase in pressure, a safety valve blew on a vent pipe, and about 2 kg of the poisonous contaminant 2,3,7,8-tetrachlorodibenzo-*p*-dioxin (TCDD) was accidentally produced and discharged to the atmosphere. A "death cloud" moved towards Milan. Eventually, 750 people were evacuated, and the zone around the plant was sealed off and surrounded by barbed wire. In a larger zone the inhabitants were given medical checks, and there was a ban on the sale and consumption of local foodstuffs.

TCDD is one of the most poisonous substances known. It is said to damage the liver, the kidneys, and the heart and cause depression, loss of memory, certain forms of cancer, and damage to the unborn.

In the aftermath of Seveso, however, only one person died of liver cancer, and no connection with TCDD was ever shown. There was a certain amount of skin disease (chloracne), and three children may have continuing moderate scarring. No employees were taken ill, and no injury to internal organs was found. All pregnant women have by now been delivered or had abortions, and the incidence of miscarriage or malformation has apparently been normal although some of the defects were of the type associated with dioxin and there have also been suggestions that the data were incorrectly reported. Even allowing for these allegations and although it is impossible to discount more subtle long term injury totally, the headline that appeared in a London so-called "quality" newspaper, "A MILLION TIMES WORSE THAN THALIDOMIDE," must have been something of an overstatement.

The Icmesa plant belonged to Givaudan, a subsidiary of Hoffman-LaRoche. It produced 2,4,5-trichlorophenol. Chlorination of benzene gives, among other things, a mixture of 1,2,4,5, and 1,2,3,4-tetrachloro-benzene, which can be separated by continuous crystallization (Fig. 6.4). The 1,2,4,5-isomer is then hydrolyzed with sodium hydroxide to 2,4,5-tri-chlorophenol, which is coupled with formaldehyde to give the general antiseptic, hexachlorophene. It may also be reacted as its sodium salt with chloroacetic acid to give 2,4,5-trichlorophenoxyacetic acid, the herbicide that gained a sinister reputation in the Vietnam War as a defoliant. This connection helped deepen the panic in Seveso at the time of the accident. The TCDD, however, was formed during the hydrolysis of tetrachloro-benzene to trichlorophenol so the connection is irrelevant. The reaction was carried out in ethylene glycol known to be a safer solvent than methanol. The reactor was heated indirectly by superheated steam to make sure that untoward temperature rises could not occur. The reaction system has been shown not to "run away" below 230°C, and the final recorded temperature on July 10, 1976 was 158°C. Two years after the accident investigators have still failed to find the cause of the explosion or the reason for the formation of TCDD.

In spite of this, scientists appointed to conduct a judicial inquiry into the accident were sufficiently upset by lack of safety equipment,

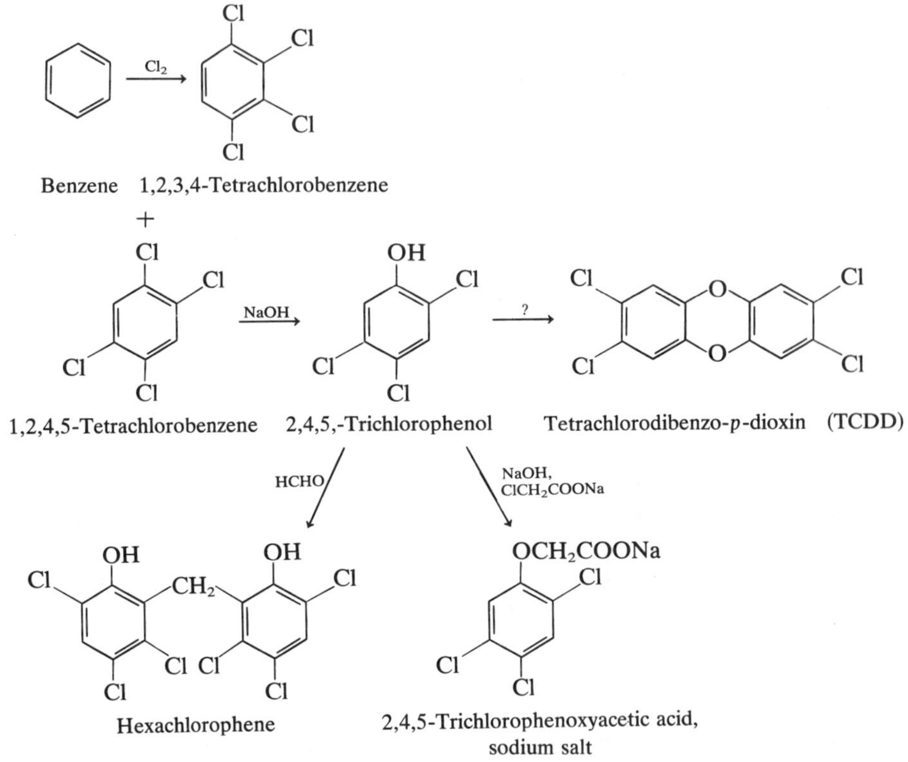

Figure 6.4   Formation of tetrachlorodibenzo-p-dioxin.

emergency systems, and adequately trained workers to place responsibility for the accident on Icmesa and Givaudan. This confirmed the findings of a nonjudicial investigation commissioned by the Italian Parliament, which resulted in charges against the companies and certain public officials.

Although the number of people who died as a result of the Seveso incident was zero, it was still an accident that might have had terrible results, and it produced in its wake a series of medical reports documenting the aftereffects of apparently minor incidents involving TCDD and demonstrating the long term potency of TCDD contamination. It was alleged by Givaudan's critics that the plant had been located in Italy because Italian regulations were slack. Previous plants operated by other companies in the United Kingdom and Germany had been permanently closed some years before because of chloracne among the process staff.

Many smaller scale incidents had been conveniently forgotten until the bigger incident called attention to them. The panic engendered by the accident is wholly understandable even if the press hysteria and government confusion tended to veil the real problems.

6.4.4 THE HEALTH OF THE STRATOSPHERE

The surface of the earth is protected from ultraviolet radiation from the sun with a wavelength below about 360 nm by the so-called ozone layer in the stratosphere about 20–40 km above the ground. The layer is maintained by a photochemical equilibrium between oxygen and ozone:

$$O_2 + h\nu \xrightarrow{<242.4\,nm} O + O \qquad O + O_3 \longrightarrow 2O_2$$

$$O + O_2 + M \longrightarrow O_3 + M \qquad O + O + M \longrightarrow O_2 + M$$

$$O_3 + h\nu \xrightarrow{<350\,nm} O_2 + O$$

(M is any inert molecule that carries away some of the heat of reaction.) It is the ozone that absorbs at the longer wavelengths and that is not present in the lower atmosphere. Any reduction in the ozone concentration would allow more ultraviolet light to reach objects on earth, and this would lead to more skin cancer among sunbathers. If the reduction were large there would be more serious damage to everyone. A 50% reduction in ozone would lead to a 130–150% increase in the dangerous ultraviolet wavelengths between 230 and 320 nm. A 10% reduction would lead to a 20–25% ultraviolet radiation increase.

Unlike other aircraft, supersonic jets such as the Anglo-French Concorde cruise in the stratosphere, and in 1970 it was pointed out that engine exhausts from 100 such aircraft would increase the stratosphere's natural uptake of water by one-third, and this could reduce the sunlight reaching the earth. Alternatively, oxides of nitrogen in the exhaust gases could reduce the concentration of ozone by way of free radical chain reactions such as

$$NO + O_3 \longrightarrow NO_2 + O_2$$

$$NO_2 + O \longrightarrow NO + O_2$$

$$O_3 + O \longrightarrow 2O_2$$

and this would increase the amount of sunlight reaching the earth.

The whole problem was much more complex than this, however, and the rate constants of hundreds of reactions were determined as part of a $20 million program supervised by the Climatic Impact Committee of the

US National Academy of Science and financed by the US Department of Transportation. The program produced work for many kineticists, finances for many university departments, and much fascinating scientific information. It may well have been justified on these grounds, but it is difficult to sympathize with the environmental issue. The stratosphere is known to be stable in that it survived the injection of vast amounts of material as a result of the Krakatoa earthquake (August 27, 1883), which was estimated to have had 100 times the power of the greatest H-bomb test detonation. Furthermore, any changes in the stratosphere could have been easily detected during the operation of the first few stratospheric flights, and the results would have been far more reliable than those determined by atmospheric modeling. The results of the program suggested that there was no danger from the flights.

The revision of rate constants brought about by the research, however, suggested that the ozone layer might be in danger from fluorocarbons used as propellants in aerosols. Two of them are predominant—trichloro-monofluoromethane CCl_3F, known as P11, and dichlorodifluoromethane CCl_2F_2, known as P12. These materials are gases but can be liquefied under slight pressure; hence their use in aerosols. They are chemically inert and have no natural "sinks." Once discharged into the atmosphere they appear to stay there, and it is claimed that the levels of about 50 parts per trillion that have been detected are what would have been expected if all the fluorocarbons used in aerosols until now had in fact been mixed into the atmosphere and had not been destroyed, dissolved by the sea, or otherwise depleted.

P11 and P12 are transparent to the wavelengths of ultraviolet light above 286 nm, which are the only ones reaching the troposphere, the atmospheric layer close to the earth. But if they are transported to the stratosphere they will be exposed to radiation of wavelength less than 230 nm, which they absorb strongly to give chlorine atoms:

$$CCl_2F_2 \xrightarrow{h\nu} CClF_2 + Cl$$
$$CCl_3F \xrightarrow{h\nu} CCl_2F + Cl$$

The chlorine atoms will then destroy ozone by way of a chain reaction in which one atom of chlorine might cause decomposition of thousands or even millions of ozone molecules:

$$Cl + O_3 \longrightarrow ClO + O_2$$
$$ClO + O \longrightarrow Cl + O_2$$
$$O_3 + O \longrightarrow 2O_2$$

The chain reaction is inhibited by the reactions

$$ClO + NO \longrightarrow Cl + NO_2$$
$$NO_2 \xrightarrow{hv} NO + O$$

These have the effect of replacing ClO molecules by chlorine atoms, the NO_2 being mainly photolyzed back to NO. It is also inhibited by the reaction $Cl + CH_4$ (or other hydrocarbons) $\rightarrow HCl + CH_3$ (or other alkyl radicals), which removes Cl atoms. On the other hand, ozone removal is promoted by $HCl + OH \rightarrow Cl + H_2O$, which adds to chlorine atoms as does a minor reaction:

$$ClO + NO_2 + M \longrightarrow ClONO_2 + M$$

Chlorine
nitrate

$$\downarrow hv$$

Various fragments + Cl atoms

It was pointed out that transport of fluorocarbons to the stratosphere would be a slow business and that by the time it could be shown that ozone was indeed being depleted the reservoir of fluorocarbons in the troposphere would be too great for anything to be done about it. Although the danger of fluorocarbons to the ozone layer could not be conclusively proved, the risk was too great to take. The disappearance of fluorocarbons and, for that matter, aerosols would cause no one any great inconvenience, and P11 and P12 should be phased out as quickly as possible.

The United States took this argument seriously. The large scale production of chlorofluorocarbons was cut off in October 1978, and sales of aerosols containing them (except for 2–3% of "essential" uses) by April 1979.

The United Kingdom and many other countries were more sanguine, and at the time of this writing still permit the products. Only Sweden has so far followed the example of the United States. Nonetheless, the United States is the greatest user of aerosols so its action will reduce dramatically the discharge of fluorocarbons into the atmosphere, even on a world basis.

6.4.5 ASBESTOS

Asbestos is magnesium silicate mixed with calcium and iron silicates in varying amounts. It is fibrous and noncombustible and is used in the production of many fireproof and insulating materials. It has been known for many years that inhalation of asbestos dust leads to shortness of breath, difficulty in breathing, and premature death. It also can lead to pleural mesothelioma, a rare form of lung cancer that appears to be caused exclusively by asbestos dust.

The interval between exposure to the dust and onset of the disease can be prolonged. In a classic case a 55-year-old woman developed pleural mesothelioma and was thought never to have been near asbestos dust. She had been born in the region of the Cape asbestos fields in South Africa but had left when she was five. It transpired that she had briefly attended school and on the way home had slid down an asbestos dump. Two of the schoolmates who had played with her on the dump also died from the same cancer. In her case there had been an interval of 50 years between exposure and effect.

Asbestos is a material that has attracted relatively little attention from environmentalists and is one of the more deadly. Many of the countries where it is mined have primitive regulations for safety at work. Many of the people who mine it are too poor to bother about long term consequences. Unlike a chemical plant where equipment can be totally enclosed, mines are difficult places to render safe, and the mining industry is much more dangerous than the chemical industry. The problem is accentuated by the fact that asbestos is a familiar material, and people tend to be philosophical about its dangers.

6.4.6 THE ACRYLONITRILE COPOLYMER BOTTLE

Glass has a number of advantages as raw material for bottles. It is transparent and can be easily cleaned, although it cannot be heat sterilized. It does not impart a flavor to beverages contained in it for long periods and is nontoxic. On the other hand, it has a number of drawbacks. Its manufacture is energy intensive, it is fragile and has a high specific gravity. Glass bottles are heavy and may break when transported.

To replace the glass bottle with a plastic bottle has been one of the aims of the plastics industry almost since its birth. Opaque and translucent bottles presented little problem and, polyolefins and PVC are widely used in such applications. Clarity was more of a problem. Péchiney-St. Gobain developed a transparent bulk polymerized PVC bottle in the late 1960s, but it tends to be brittle when squeezed. It is also difficult to compound PVC with ingredients acceptable for contact with food. Poly(4-methylpentene-1), seen at one time as a possible substitute for glass, turned out to be more expensive than expected.

The glamor market, however, is for a clear container for packaging carbonated beverages such as soft drinks and beer. By the beginning of

1977 Monsanto thought they had found the answer with a blow molded bottle that was simultaneously stretched in two directions to give a biaxially oriented polymer. The resin could be either poly(ethylene terephthalate) or the much cheaper acrylonitrile copolymer resins. In 1977 Monsanto and other companies expected to consume more than 10,000 tons of nitrile resin to make soft drink bottles. Early in 1977, however, the US Manufacturing Association reported that animal tests had shown acrylonitrile to be potentially carcinogenic. In May of that year duPont announced that studies carried out at its Camden, South Carolina acrylic fiber plant had shown "excess cancer incidence and cancer mortality as compared with the company and national experience." This amounted to 16 cancer cases among a test group compared with 5.8 that would have been expected based on company cancer rates and 6.9 considering national cancer rates.

The US Occupational Safety and Health Administration introduced an emergency exposure limit of 2 ppm over an eight-hour period. The commissioner for the Food and Drug Administration ruled that acrylonitrile was a food additive and was liable to migrate into beverages even though the most sensitive available tests (10 ppb) had failed to detect such migration. It was agreed that acrylonitrile was a carcinogen when administered in large doses to test animals. Consequently he banned the container. US law apparently does not state that migration of a material into food needs to be demonstrated but only that it might reasonably be expected to do so. As it was impossible to show that traces of acrylonitrile monomer below the limits of detection did not remain in the polymer, this condition was fulfilled.

Poly(ethylene terephthalate) bottles have appeared on the market but as they are more expensive they have had a smaller impact.

6.4.7 THE "ACCEPTABLE" RISK

The various disasters discussed and problems raised in the preceding pages have virtually nothing in common. While we all have 20/20 vision in hindsight, it would be a bold person who would claim to have foreseen any of the various disasters in terms any more specific than a general blanket disapproval of the chemical industry. The chemical industry, moreover, would claim that in the vast majority of cases it had cooperated closely with the government both to monitor problems that might arise and to mitigate the consequences of accidents once they had happened. It would also point out that it has a far better safety record than traditional industries such as construction, mining, or deep sea fishing. It is not rational to complain more about a person who dies of pleural mesothelioma 50 years after working in an asbestos mine than one does about another person who is killed instantly by a rock fall in a coal mine.

The industry is correct in this view and is right to protest about double standards. The fact remains that people are not rational and fear hidden dangers that may manifest themselves only in the long term more than they fear direct and obvious dangers like rock falls or drowning. Once people have survived the direct dangers they feel the slate is wiped clean and they can resume their daily lives uninhibited by fears of slow but insidious poisoning. Their fears are exacerbated by the media, which find a chemical industry accident more newsworthy than an old-fashioned drowning.

To counter this view chemists have come up with the idea of the "acceptable risk." All human activities involve risk. Walking, riding, eating, and sleeping can lead to disaster if the chain of coincidence stretches out its hand. Accident statistics show that the most dangerous place is the home. People nonetheless continue these activities (or stay at home) because they have weighed the risk of so doing on the one hand and the need for it on the other, and have concluded that the risk is acceptable. Which risks are indeed acceptable might vary from person to person and situation to situation. An adult would cross a busy street but an infant would be ill-advised to try to do so. A strong swimmer would dive into the deep end of a swimming pool without hesitation while a sensible nonswimmer would remain on the edge. A patient suffering from terminal cancer might well be given pain-killing morphine by a doctor who would certainly not prescribe it for a severe headache. Sodium is explosive and chlorine poisonous, but few would hesitate to sprinkle sodium chloride on their food for fear of residual amounts of these elements.

If the idea of "acceptable risk" is to be useful we need a standard by which different risks can be assessed so that we can devote resources to diminishing the major risks and not waste them on minor ones. Risks are difficult to quantify, and numbers attached to them should be taken as precise estimates only if the methodology is understood and the assumptions clearly stated. Nonetheless, the quantifiable risk can be illustrated by Table 6.2, which lists 15 activities each of which carry a one-in-a-million chance of death. Risks smaller than this are usually taken freely; even much larger ones are accepted as inevitable. A normal day's movement in the United Kingdom carries a one-in-two-million chance of a fatal road accident, but staying at home is even more dangerous. The most dangerous activity in Table 6.2, and the one which brings the fewest obvious benefits to its practitioners, is cigarette smoking. Few of the risks are due to the incompetence or malevolence of big business.

Table 6.2   Risks of One-in-a-Million[a]

How Much	Of What	Gives a One-in-a-Million Risk of
1.4	Cigarettes	Cancer, heart disease
2 months	Of living with cigarette smoker	Cancer, heart disease
0.5 litre	Wine	Cirrhosis of the liver
10 tablespoons	Peanut butter	Liver cancer caused by aflatoxim
1 gallon	Miami drinking water	Cancer caused by chloroform
30 cans	Diet soda	Cancer caused by saccharin
2 months	In stone or brick building	Cancer caused by radioactivity
2 months	Visit to Denver	Cancer caused by cosmic rays
3000 miles	Jet flying	Cancer caused by cosmic rays
1 month	Diagnostic x-rays	Cancer
20 years	Living near PVC plant	Cancer caused by vinyl chloride
1 day	In New York or Boston	Death by air pollution
3 hours	In coal mine	Accident
1 hour	In coal mine	Black lung disease
1000 miles	Jet flying	Accident

[a] Compiled by R. Wilson of Harvard Energy and Environmental Policy Center.

The idea of balancing risk against benefit is a sensible one and is accepted in most countries and enshrined in most laws. It is ignored however in the 1958 Delaney Amendment to the US Food, Drug, and Cosmetic Act. This prohibits the use of any chemical substance in any amount as a food additive if that substance has been found by "appropriate" tests to induce cancer in man or laboratory animals. The amendment was originally designed to protect consumers from substances intentionally added to foods, but modern analytical methods are able to detect in foods microscopic quantities of materials that when given to rats in enormous doses induce cancer. These materials may have found their way accidentally into the food or have always been there. If analytical methods continue to improve it may soon be possible to show the existence of carcinogens or toxic materials in virtually every foodstuff.

It is illogical that cyclamates (artificial sweeteners classified as chemical substances) are banned while cigarettes, which are very much more dangerous, count as "natural" and are therefore legal. It is significant that evidence of the kind used to ban cyclamates was later brought against saccharin. Public reaction to the attempt to ban the second major artificial sweetener was so hostile that a moratorium was placed on the ban for 18 months, which was recently extended (Part II, Sec. 13.3.3). At some stage in the next few years the interpretation of the Delaney Amendment will have to be modified to permit a risk/benefit balance to be examined, and similar comments apply to other areas of environmental control. In some areas control may need to be tightened; in others it may need relaxation. In the debate that surrounds the issue it is important for chemists to be well informed and to be able to assess areas where industry could usefully aim more resources and other areas where concern is scarcely justified.

The eventual legislation on all the problems loosely described as environmental will have an important influence on the future of the chemical industry. The extreme environmentalists are unlikely to win a complete victory if only because the US populace is wedded to a high standard of material comfort. Equally, the chemical industry will never be above government control as were the "robber barons" of the nineteenth century. Nor would it want to be; it has shown willingness to cooperate with government agencies at every level. A compromise will undoubtedly be reached, but whether it will be a rational one in which resources are concentrated on the real problems depends on the integrity and good sense of the various protagonists.

6.5 OVERCAPACITY AND SHORTAGE OF CAPITAL

The chemical industry has a tendency toward overcapacity. It is capital intensive, uses high technology, and has a good record of growth, all of which make it attractive to investors. Its growth record means furthermore that chemical companies themselves have had steadily increasing cash flows to invest, and they were naturally inclined to invest these in chemical enterprises. For the first half of the twentieth century the tendency was limited on occasions either by government action (UK "key industry" protection of the infant Imperial Chemical Industries from foreign competition in the 1920s; the "American Selling Price" system under which tariffs were levied on the price at which US manufacturers sold the material being imported) or by cartels and market-sharing agreements under which the strong flourished and the weak survived.

After World War II, government pressure, especially in the United States destroyed the cartel system, and with the petrochemical revolution the major chemical companies installed new capacity at an unprecedented rate. At least four good reasons for installing more capacity for a chemical than the market required follow:

1 The additional capacity would be needed in a year or two because of the rapid growth of the market. A competitive attitude might make possible increase in market share and export of surplus

production. The argument is partly valid, but not everyone can simultaneously increase their market share nor can all countries export more than they import.

2 At a time of inflation it is better to build now rather than wait until next year when the market will have increased, because next year the plant will cost more.

3 One might well have some new process to commercialize. The idea, of course, is to drive one's competitors with obsolescent processes out of business, but this does not always happen. In a capital intensive industry there is a large gap between the price at which one covers all one's costs including capital costs and the price at which one can pay variable costs—raw materials, services, and labor. As long as one can do the latter it is worth staying in the market.

4 At a time when economies of scale were particularly attractive it was tempting to build too large a plant and once again trust to luck and exports to keep it working at capacity.

These reasons were cogent, and during the 1950s and 1960s expansion was fueled not only by cheap oil but also by the rising cash flow from the growth of the industry and the willingness of the investor to accept a low return on captial provided the cash flow was plowed into further expansion. Consequently the chemical industry experienced long periods of overcapacity and "profitless prosperity." Some of the luckier, better managed or more prudent companies thrived, and some products were consistently in short supply. But as a whole the beneficiaries of the great boom in heavy organic chemicals and polymers were the downstream users who were able to obtain their raw materials at steadily decreasing real prices.

The depression of 1973 and the slump of 1975 led to pessimism in many chemical companies and the cancellation of many investment projects. On classical economic theory this is just what ought to happen. Only when the surplus capacity installed in the days of optimism is taken up will prices rise sufficiently to induce further investment. Thus the overcapacity will be self-correcting.

This seems unlikely because it assumes a perfect market without government intervention. In fact, the situation is far from perfect and varies from country to country. It is summarized in the next section for selected areas of the world.

Figure 6.5a shows the total world chemical production for 1974 (1975 figures are similar) and the forecast figures for 1985. Figure 6.5b illustrates the same figures expressed as output per capita. Ninety percent of the world chemical production comes from four blocs—North America

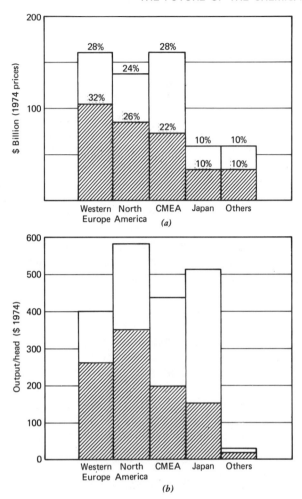

Figure 6.5 (*a*) World chemical production (excluding China). (*b*) World chemical production per capita population (excluding China). Actual figures are given for 1974 in the shaded areas; estimated figures are shown for 1985.

(United States and Canada), Western Europe, Council for Mutual Economic Assistance (CMEA or COMECON, comprising the USSR and most of Eastern Europe), and Japan. The growth rates implied for the developed nations by the forecast in Figure 6.5 seem rather high, and it is possible that the development of indigenous chemical industries by the OPEC (Organization of Petroleum Exporting Countries) nations will increase the share of "others" above the 10% that is forecast. Nonetheless, the share of world chemical production of the four major blocs is unlikely to drop below 85% by 1985, and trade, investment, and transfer of technology between them will be the dominant feature of the world chemical scene in the medium term future. There is an enormous gap between them and the "others." The other noteworthy feature of Figure 6.5 is that the CMEA chemical industry is expanding rapidly and is expected to overtake North America in total production (although not in production per capita) in the next decade.

Figure 6.6 shows the trade between the blocs. Western Europe exports and imports more chemicals than anyone else but imports much less than it exports—$5700 million as against $15,000 million. Many of its exports

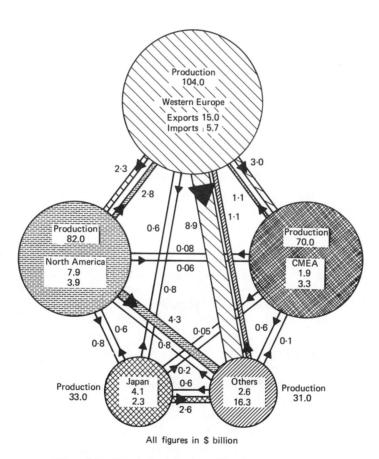

All figures in $ billion

Figure 6.6 Chemical production of interbloc trade, 1974.

go to "others," and its trade with North America and Japan is essentially in balance.

North America is the next largest importer and exporter and has a positive trade balance due largely to its net exports to "others." Japan is also a net exporter while CMEA and "others" are net importers, both blocs drawing heavily on Western Europe.

The four major blocs all import between 5 and 7% of their consumption of chemicals, but their exports range from 3% of production for CMEA through 10% for North America and 12% for Japan up to 14% for Western Europe. Even this large figure neglects the large intra-Western European chemical trade which in a sense counts as imports and exports. A company such as ICI sells about half its output outside the United Kingdom.

This leads us to the problems affecting the industries in the various blocs.

6.5.1   UNITED STATES OF AMERICA

The United States constitutes a large homogeneous market without significant national boundaries. Its tariffs protect it from excessive foreign competition, and it is distant from its competitors. Its chemical industry had cheap, abundant raw materials until recently and to some extent still has. It is wholly in private hands, and equity and successful risk taking are well rewarded. However, there are increased antitrust pressures, problems with overenthusiastic environmental pressure groups, energy problems, and inflation.

Tariff and nontariff restrictions to imports are such that if a foreign company wishes to enter the US market in a big way, it is almost forced to set up a manufacturing facility in the United States even when transportation costs would make direct trade practical. This factor, combined with the current (1978) relative cheapness of the dollar, means that foreign investors are eager to invest in the United States.

Politically and economically the United States is the strongest country in the world. While it cannot view with equanimity the extension of Communist influence in Europe, Africa, or the Middle East, it can afford to view such prospects with greater detachment than those living nearer the trouble spots. If the rest of the world would go away, North America could survive. The other side of coin is that the United States is the hope of the free world, and minor fluctuations of US policy can spell catastrophe for those living in trouble spots. In addition, no energy conservation policy has a hope of success unless adopted by the United States because the United States consumes such a large proportion of world energy. Compared even with other rich countries such as Sweden, the United States is profligate with energy and shows little sign of changing.*

6.5.2 WESTERN EUROPE

With the enlargement of the European Economic Community (EEC) and the elimination of tariff barriers between the EEC and the European Free Trade Area (EFTA) it should be possible in theory to regard Western Europe as a strong, cohesive, homogeneous block. No view could be more mistaken. Within Western Europe national economies dominate, national boundaries strongly influence investment, and economic chauvinism means that pricing policies are weak. Countries typically have "mixed" economies, and there is constant pressure on the national chemical industries to build new plants whether or not they can sell the product. In the past, for example, the Italian government poured in public funds to restore the economy of the Mezzogiorno region, and this was spent on the overbuilding of petrochemical and fiber plants. Ironically they provided very little employment because such plants are capital, not labor, intensive. Similarly, the UK government has suggested to the British chemical industry that it should build four or five crackers in Scotland, each to produce 1 billion lb year^{-1} of ethylene from North Sea oil and natural gas liquids. The industry's own market estimates are that one or conceivably two new crackers would keep the market well supplied for the foreseeable future.

Already built is the huge new Norsk Hydro complex at Rafnes, which cracks natural gas liquids from the Ekofisk field. Enthusiastic statements and participation by the Norwegian government cannot conceal that the venture is likely to be a financial disappointment.

In "mixed" economies manufacturing companies are too often used as agents of government policy and are encouraged to ignore profitability. Unnecessary personnel are hired to reduce unemployment; too much capacity is installed, and the products are sold at depressed prices. As the companies approach bankruptcy, left wing parties claim that capitalism has failed and propose nationalization or greater state control as the price for further government aid. Companies that resist overbuilding are accused of intransigence and willful refusal to participate in the government's messianic vision.

---

* We wrote the above in 1977 and since then United States industry has made impressive strides in conserving energy. The general public, however, remains profligate, and firm action by the government seems unlikely. We would very much like to be proved wrong in this gloomy assessment.

In addition to this factor, the high level of exports tends to reduce profits. Transport costs have to be paid and tariffs absorbed. Prices may have to be trimmed for a company to break into a foreign market at all.

Thus the West European chemical industry is not as profitable as it should be, and its prices are generally too low to permit profitable new investment. It is the least cohesive of all the blocs, has the least protected and most heterogeneous market, and is the biggest importer and the biggest exporter. In short it is the most vulnerable.

6.5.3 USSR AND EASTERN EUROPE (CMEA)

The chemical industries of the CMEA are growing rapidly. The chemical industry has the advantage of being capital intensive but not prohibitively so, and the CMEA rulers see the industry as satisfying some of the pent-up consumer demand. They are able to dictate what is sold as well as what is made since their markets are totally protected. Capital investment comes out of state funds, and balance sheets are of little importance.

Much of the technology, particularly for the more sophisticated organic chemicals, needs to be bought, and the CMEA has turned to Western Europe. There, no one company or country has a monopoly whereas the CMEA countries are monopoly buyers. Because of the world recession, West Europan financial institutions were holding excess cash, and governments faced with unemployment in their machinery and engineering industries rushed to finance deals with the East on terms considerably more lenient than would be granted to a home-based borrower. Because the CMEA countries were short of foreign exchange many of the "technology transfer" agreements were financed on a payback basis, that is repayment not in cash but in quantities of the product of the eventual plant.

These arrangements carry obvious dangers to Western European chemical companies. First, the influx of large quantities of a product "paid" to a West European contractor for construction of a plant in Eastern Europe could add to prevailing oversupply in Western Europe and depress prices. Second, the cheap credit granted to CMEA countries means that they may compete with home-based companies on unequal terms. Third, the USSR, given the size of its investments in chemicals, could undertake to deliver through payback deals quantities of a product as big as the whole international trade in it. It has been suggested that this is almost the case with respect to ammonia. Having occupied the whole of the international market the USSR could withdraw from it suddenly for economic or political reasons, and the West could do nothing. Fourth, the USSR and other communist countries, notably Cuba, themselves provide low interest loans and these go to so-called liberation movements or "radical" governments in the form of arms and aid designed to cause maximum disruption in the free world. Thus the West European taxpayer is indirectly subsidizing world terrorist movements and at the same time undermining his own indigeneous chemical industry.

From the Communist point of view, of course, these difficulties represent the internal contradictions of capitalism, and they are delighted to be able to buy technology and access to West European markets so cheaply. "Detente" has produced rich dividends for the USSR, and since the West is in no position to insist on the implementation of that section of the Helsinki Agreement dealing with human rights there have been few adverse consequences.

In the long run, and probably more slowly than CMEA planners think, their domestic markets will expand and absorb the production, but unless West European policy changes there will be further arrangements of the same kind. Eventually the communist bloc will "acquire" Western technology just as the Japanese acquired it in the 1960s, and they might be able to make the impact in the future that the Japanese chemical industry made in the 1960s.

On a more optimistic view, Western Europe might be able to stay ahead in technology, and CMEA and Third World domestic markets might expand more rapidly than expected.

6.5.4 JAPAN

The Japanese economic miracle was the economic phenomenon of the 1960s. Lacking indigenous energy resources and homegrown technology Japan accomplished it by import of both. All the developed countries have felt the impact of the excellent Japanese goods far removed from the shoddy imitations of Western goods produced with cheap labor before World War II. The shock of 1973 upset the boom, and the Japanese chemical industry is suffering from low profitability and growth. Like the West Europeans, the Japanese are looking abroad for chemical plant investments and have initiated projects in Singapore, Iran, and the Persian Gulf. In the end this is bound to increase overcapacity just as Western European transfers of technology to Eastern Europe will do. The Japanese are already complaining about the Taiwanese and South Koreans selling cheap synthetic fibers in the traditional Japanese markets of Southeast Asia, but it was the Japanese themselves who designed and built most of the fiber plants and, many years previously, the United States and Western Europe who transferred the technology to Japan.

Japan is the dominant producer of chemicals in the Pacific, has high tariff barriers, and has a high work ethic. On the other hand, she has no indigenous raw materials and is second only to Western Europe in her reliance on exports. She appears vulnerable but might ultimately emerge as the leading partner in some short of Southeast Asian Common Market.

6.5.5 THE MEMBERS OF OPEC

The Organization of Petroleum Exporting Countries is a cartel that controls the world's long term reserves of petroleum. In 1973 it had thirteen members—Algeria, Ecuador, Gabon (associate), Indonesia, Iran, Kuwait, Libya, Nigeria, Qatar, Saudi Arabia, United Arab Emirates, and Venezuela. Some of these countries are traditional right wing absolute monarchies (Saudi Arabia, Qatar, United Arab Emirates) while others are "radical" left wing oligarchies (Libya, Algeria). Some are sparsely populated deserts (Saudi Arabia) while others (Iran, Nigeria) have sizable populations. That they succeeded in imposing even a reduction of petroleum supplies to the West in 1973 was a surprise, and there is a question even now as to the degree of cohesion between the members. Nigeria wants high prices and large sales to finance her ambitious development plans while others are less concerned with money and output and more concerned with politics and religion.

Nonetheless, even the threat of an oil embargo cannot easily be forgotten by Japan or Western Europe which, if we except North Sea oil, are wholly dependent on imports. US imports of Saudi Arabian crude have escalated alarmingly in recent years, and unless she takes action to reduce energy demand she too will find herself embarrassingly dependent on a handful of politically unstable dictatorships.

From the point of view of the chemical industries in Western Europe, North America, and Japan, however, the members of OPEC provide another problem. They are now building or planning to build massive petrochemical industries downstream of their refining capacity. The output of these plants will be far greater than the countries concerned (with the possible exception of Iran) will be able to absorb through their domestic markets for many years to come. They too will be looking for markets abroad, primarily in Western Europe. The capital for such plants is readily available from oil revenues, and the pricing of the products can be arbitrary since raw material and energy requirements can be costed in at a small or zero value.

The construction of petrochemical plants in the less developed countries of OPEC presents formidable problems ranging from the lack of a technological infrastructure in their societies to such apparent trivia as lack of fresh water for cooling purposes in many of the countries. Many of the more bizarre plans for the Persian Gulf sheikdoms put forward in the heady days of 1974 have been quietly dropped, and the pace of construction of the remaining projects is slow. By 1985 their output is likely to be only a few percent of world production, but it could grow rapidly after that. Western Europe has so far resisted requests to lower tariffs on or provide quotas for OPEC chemicals, but OPEC could some day bring indirect pressure by threats to withhold vital petroleum supplies.

One pattern that may overcome some of the problems faced by OPEC is collaboration with Western or Japanese companies that would provide technology and marketing in return for guaranteed crude oil supplies. A consortium consisting of Saudi Arabia Basic Industries Corporation, Celanese Chemical, and Texas Eastern Arabia has recently announced plans to build a 2000 ton per day methanol plant in Saudi Arabia. This is equivalent to the size of the largest single stream organic chemical plant existing anywhere in the world. While this collaboration may benefit the consortium, the fact that the Saudi Arabian methanol market is nil raises yet again the specter of world overcapacity.

6.5.6 THIRD WORLD COUNTRIES OUTSIDE OPEC

Finally we come to the world's poorest countries—those less developed countries (LDC's) that lack oil. While many of them possess other important raw materials (e.g., copper in Zambia, bauxite in Guinea, Jamaica, Guyana, and Surinam) efforts to organize OPEC-like cartels have failed. The LDC's furthermore have been worse hit than anyone else by the quintupling of petroleum prices. They face a dismal future in which the world trade recession makes it increasingly difficult for them to earn foreign exchange or solicit foreign aid, and meanwhile their populations continue to grow typically at 3% per year. As fuel imports increase their indebtedness, more and more of their foreign exchange goes to service past borrowings, and they become less and less credit worthy.

The problems of the LDC's have been compounded by bad management and poor leadership. The majority have opted for showy development projects and industrialization that have caused migration of peasants into the towns and exacerbated social problems. Money would surely have been better spent on improvements in agriculture, although LDC leaders might well reply that nations that have lavished funds on moon landings and supersonic passenger aircraft have no right to criticize others for their concern with prestige.

Few of the LDC's have chemical industries of any size, although a number of them manufacture fertilizers. Others such as South Korea,

Taiwan, Singapore and Hong Kong have bought know-how on synthetic fibers. They have taken advantage of the fact that textiles are labor intensive, and the cost of fiber is a relatively small item in the cost of a garment. Thus they have exported a range of textiles, some of excellent quality, to Western Europe, North America, and Japan. As usual, Western Europe has borne the brunt, but throughout the world textile industry there is massive overcapacity and severe unemployment. On the other hand, it is difficult to know whether or not the LDC's would have been better off had they not entered the textile industry but spent the money on something else. From a common sense point of view they would appear to do better to enter the fertilizer industry, but this would have the drawback of not generating foreign exchange.

Whatever the moral obligations of the West (or the Communists or the OPEC) toward the LDC's there is little economic incentive for foreign investment, technology transfer, or even foreign aid.

6.5.7 AVAILABILITY OF CAPITAL

On the classical economic model, money would be invested in the chemical industry in those countries where the chemical industry had a "comparative advantage," that is, where the inhabitants were relatively better at making chemicals than they were at doing other things. To take a homely example, it would be more sensible to invest in Scotch whisky distilleries in Scotland and champagne-making facilities in France than vice versa.

This simple picture is distorted by the fact that in few of the world's countries are decisions taken on purely economic grounds. A North American firm considering investment in chemicals will certainly have to consider profitability, and so will firms in West Germany and Japan. In the United Kingdom, market discipline is relaxed, though only slightly, by the possibility of development grants and money from the National Enterprise Board, and similar situations exist in Italy and France. In the CMEA countries investment decisions are taken on political not economic grounds, and, although the political grounds may be different, the same is true for many LDC's. The members of OPEC, of course, have huge amounts of capital readily available.

If North America, Western Europe, and Japan wish to remain in the business of producing large tonnage organic chemicals they will have to consider problems of availability of capital. There are many producers who are eager to get a foothold in the market and who are indifferent to profits and, accordingly, to overcapacity and low profitability. Tariffs may help to some extent, but they are rightly seen as unfortunate barriers to world trade. But if profitability is low there is no incentive for profit-oriented companies to invest, and the chemical industry will gradually drift away from its present locations.

The problem is exacerbated by a dramatic rise in the cost of plants. The No. 3 naphtha cracker at Grangemouth in the United Kingdom, commissioned by Distillers Company/British Petroleum in 1960, produced 85,000 tons per year of ethylene and cost £6 million. The No. 4 cracker, commissioned in 1968, produced 250,000 tons per year and cost £8 million. The joint Imperial Chemical Industries/British Petroleum cracker currently being commissioned at Wilton will produce 500,000 tons per year and will cost £150 million from which, to be fair, one should subtract £11 million for the butadiene unit and £16 million for the pipeline to Grangemouth.*

There are many reasons for the price increase. Part of it is due to inflation and part to higher environmental standards. Depending on the size of site and degree of purification required, up to 13% of total planned investment may be required to provide facilities for waste water treatment and up to 25% for a total effluent package. Whatever the reasons it is increasingly difficult for a single chemical company to finance an ethylene cracker, let alone a whole petrochemical complex. The limit to future expansion of the chemical industry in Western Europe and perhaps also in Japan and the United States may well be cash flow allied to low profitability.

6.5.8 A GEOGRAPHICAL SHIFT IN THE CHEMICAL INDUSTRY

We have written the preceding section (and indeed the whole of this book) from the points of view of two observers who have worked in the US and UK chemical industries and academic institutions. We are concerned both for the survival of the industry and for the survival of the academic traditions that are part of the culture of the so-called free world. Because of this, in the preceding section we have perhaps implied that the transfer of technology to countries outside the free world is something of a threat. This is true only insofar as it is used to undermine the free world in political terms. There can be no objection to technology transfer as such. Indeed, much of human progress has been due to the diffusion of new technology through different societies. The stirrup was invented in the Middle East and took several centuries to reach Britain and several more to reach America. Iron was discovered about 1500 B.C.

---

* Fluctuations in exchange rates over this period make it meaningless to express these figures in dollars.

in the Caucasus and by 1100 B.C. had reached Syria from where it spread to Egypt, Carthage, and along the Mediterranean. The Caucasians could hardly have expected to keep it to themselves.

Equally, the developed countries cannot expect to keep the chemical industry to themselves. As a technology becomes better established and better understood it will naturally move from highly developed countries to less developed ones. The process will always cause grumbling in the developed country, but in the past it has usually led to increases in prosperity for all concerned. The price of the developed country's staying ahead is the discovery of new technology on which it can base its prosperity. There is only a tenuous connection between a country's prosperity and its natural resources. The real correlation is with the education and ingenuity of its people and the social and political framework in which they work.

Technology transfer in itself is excellent. It is in any case inevitable; but even if it were not it should be encouraged. It is bound to be a feature of the next few decades, and we might well see the transfer of large sections of the chemical industry to other regions of the world. For example, the heavy organic chemicals sector might move to the members of OPEC leaving the higher technology sectors such as pharmaceuticals in the developed countries of the West. The political consequences of this are worrisome, but economically there is something to be said for it. If the developed countries cannot make a profit out of heavy chemicals and someone else wants to try, then why not let them? Why should the developed countries not buy heavy chemicals from developing countries as they now buy raw materials and concentrate their energies on the more profitable downstream sectors of the industry? The political objection is the only cogent one, and in the future we may expect to see geographical shifts in the chemical industry toward the CMEA countries, OPEC, and the less developed countries. The developed countries will try to preserve their own heavy chemical industries partly by staying ahead technologically and partly by tariff barriers to prevent "dumping." A revival of the cartel system is a possibility.

6.6 VALEDICTION

In Part I of this book we have tried to show the place occupied by the heavy organic chemicals industry in the economy of the United States and other developed nations. We have described how the majority of heavy organic chemicals are made from petroleum and natural gas and how a minority are derived from coal and agriculture. We have pointed out the convenience of petroleum and natural gas as raw materials but have also stressed the versatility of industrial organic chemistry and the opportunities it provides for obtaining chemicals from other sources were petroleum and natural gas to run out.

Finally, we have discussed the future of the chemical industry and considered certain of the technical, social, economic, and political factors that may affect it. We considered the most important to be energy and availability of raw materials, development of new catalysts, problems of toxicity, pollution, and safety at work, and political-economic problems connected with Eastern Europe/USSR and with OPEC and the Third World countries.

In 1972, a year before the 1973 economic crisis, one of us wrote, "Whichever way one looks at the data, the conclusion is inescapable; the growth of the industry will slow down." This forecast has been amply borne out. The chemical industry is approaching maturity, and its growth would naturally be expected to decline to approach that of the economy as a whole. Resources are being diverted from "growth" to environmental problems. The rising price of petroleum has increased the price of chemicals and reduced the rate at which they are substituting for traditional materials. Some of the markets for chemicals were, in any case, approaching saturation.

By the standards of the last 40 years, the world economy has slowed down sharply and perhaps permanently. The cost of energy means that the unprecedented economic growth rates of the golden fifties and sixties are gone, never to be repeated.

Growth may also be inhibited in developed countries by the setting up of chemical industries in countries that have previously been net importers of chemicals, and these countries may then wish to become exporters, with consequent world overcapacity for chemicals and low profits.

The developed nations, if they wish to preserve their privileged position, will have to do so by vigorous research programs, and from the point of view of the scientist there will be plenty to do. It is, after all, just as exciting to search for a natural product to replace a petrochemical product in a low growth chemical industry as it is to look for a new petrochemical process in an expanding industry.

The foregoing statement may seem remarkable to people conditioned by a quarter of a century of rapid growth, but other generations from Neolithic man onward would find the present day obsession with rapid economic growth even more peculiar. The trees do not, after all, grow to the sky.

The danger is not so much that economic growth will slow down (mankind lived for millennia without petroleum, and such people as Moses, Cleopatra, Christopher Columbus, Leonardo da Vinci, and

Shakespeare flourished without it) but that our social institutions will be unable to endure the frustration of the expectations of the population. Would zero economic growth lead to political instability? To violence? To nuclear war?

Technology is the handmaiden of mankind. We have described in this book some of its scope and versatility. It will do the jobs mankind wants it to do, confined only by such limitations as the second law of thermodynamics. It cannot bring us a golden age or save us from Armageddon. That can only be achieved by people working rationally together within sensible and humane social and political institutions that they have set up. The chemist has a vital role to play here, and with an understanding of technology he can help ensure that wise decisions are made and that technology is used to benefit and not to doom mankind. The future of mankind will be decided in the final analysis not by the conquest of the atom nor of outer space but in the battle for the hearts and minds of men.

To be an industrial chemist in the 1950s and 1960s was exciting. It can be just as exciting in the future, but the aim should not be 4 or 6 or 8% growth per year (although this might help) but the sort of society where "Every man shall sit under his vine and under his fig tree and none shall make him afraid."

REFERENCES AND NOTES

In this section we have confined ourselves to the medium term future and have not involved ourselves in the discussion of world dynamics. Those interested should start by reading D. H. Meadows et al., *The Limits to Growth*, Potomac Associates for the Club of Rome, Washington, DC, 1972, and follow it with the refutation in *Futures*, **5** (1973) 1 by the Science Policy Research Unit of the University of Sussex. They can then diversify into the abundant more recent discussions.

Sec. 6.1   Much of the material in this section is drawn from our own experience. Further sources are cited in B. G. Reuben and M L. Burstall, *The Chemical Economy*, Longman, London, 1974, and B. G. Reuben, *The Elizabethan Era—The Polymer Age, Chemical Engineer*, July 1977, p. 515.

Sec. 6.1.1 Although Germany was not interested in petrochemicals, she was interested in ethylene, among other things as a feedstock for ethylene glycol antifreeze for her tanks on the Russian front. German ethylene production in 1943 was 25 million lb. It was obtained mainly by hydrogenation of acetylene and dehydration of fermentation ethanol, but 3 million lb were obtained from the cracking of ethane from the Fischer-Tropsch process and from coke oven gas. The history of ethylene is excellently dealt with by H. M. Stanley in *Ethylene and its Industrial Derivatives*, S. A. Miller, Ed., Benn, London, 1969, Chap. 2.

Sec. 6.1.2 Figure 6.2c is based on data provided by Dr. M. Dilke of B.P. Chemicals international.

Sec. 6.3 Hydrogen as a fuel is advocated by J. O'M. Bockris, *Environmental Chemistry*, Plenum Press, New York, 1975, Chaps. 16–19, although he sometimes appears to lose sight of its being only an energy carrier. The question of oil and coal as sources of polymers and energy is considered by H. G. Elias, *CHEMTECH*, **5** (1975) 748 and **6** (1976) 244, and he concludes that it will be better to use coal for energy and oil for polymers. Use of plants as fuel is discussed by G. C. Szego and C. C. Kemp, *Energy Forests and Fuel Plantations*, *CHEMTECH*, **3** (1973) 275, and more recently by F. D. Lectig and D. I. H. Linzer, *Fuel Crop Breeding*, *CHEMTECH*, **8** (1978) 18.

Energy intensiveness is defined as cost of direct and indirect purchases of primary fuels divided by value of total output $\times$ 100%. The calculations were performed for the UK industry (*The Increased Cost of Energy—Implications for UK Industry*, National Economic Development Organization, HMSO, London, 1974), but the rank order will be similar for the United States.

It appears (*USI News*, *USI Chem.*) that only about 0.7% of the hydrocarbons in natural gas is removed for chemical production. Stated another way, natural gas is about 5% ethane, and only 15% of this 5% is removed for chemical purposes. If more were removed the Btu content of natural gas would be affected only slightly, and the supply picture for ethylene for chemical purposes would be much improved.

A discussion of the technology for obtaining crude oil from shale and coal has been published by D. L. Klass, *CHEMTECH*, August 1975, p. 499. The article compares the various methods for recovering oil from oil shale. The Aman process is discussed by B. G. Reuben, *Eur. Chem. News*, April 18, 1975, p. 30. The electrothermal route to phosphoric acid is discussed in *Worldwide in Phosphorus Chemicals*, Albright and Wilson, Ltd., 1 Knightsbridge Green, London SW1, England. Synthetic versus natural rubber is reviewed by P. W. Allen and L. Bateman, *Proc. ECMRA Conf.*, Aix-en-Provence, 1973, and in *Chem. Age*, May 13, 1977, p. 7. The estimates of the effect on costs of a 100% increase in the price of naphtha appeared in *Chem. Age*, June 1, 1979, p. 14.

Sec. 6.4 The best textbook we have found on this somewhat overworked topic is J. W. Moore and E. A. Moore, *Environmental Chemistry*, Academic Press, New York, 1976.

Sec. 6.4.1 *The Flixborough Disaster*, Department of Employment, HMSO, London, 1975 (Report of the Committee of Inquiry).

Sec. 6.4.2 C. and P. Revellue, *Source Book on the Environment, The Scientific Perspective*, Houghton Mifflin, Boston, 1974.

Sec. 6.4.3 *The London Times*, July 21, 1976 (first reports in the United Kingdom); *Eur. Chem. News*, August 13, 1976. The chemistry of Seveso was presented in *Eur. Chem. News*, August 13, 1976, p. 4. The term dioxin is confusingly used as a synonym for TCDD since the *Merck Index* attaches it to a quite different chemical. Atempts to discover the source of the explosion are recounted in A. Hay, "What Caused the Seveso Explosion?," *Nature*, **273**, (1978) 582. The first book to appear on the topic was J. G. Fuller, *The Poison that Fell from the Sky*,

Random House, London, 1978.

The findings of the nonjudicial inquiry were reported in *Chem. Week*, August 9, 1978, of the judicial inquiry in *Chem. Week*, February 14, 1979, and the allegations of misreporting in *Chem. Week*, February 28, 1979.

Sec. 6.4.4   The petition against fluorocarbons was reproduced in full in *CHEMTECH*, **5** (1975) 44 and contains detailed references to other work. The maintenance of the ozone layer is part of the overall chemistry of the atmosphere, an area that has received much attention in the last decade. The data here were kindly provided by Prof. F. Kaufman. A useful book for chemists is M. J. McEwan and L. F. Phillips, *The Chemistry of the Atmosphere*, Edward Arnold, London, 1975.

Sec. 6.4.6   The response of the Commissioner of Food and Drugs to Monsanto's application for approval of the acrylonitrile copolymer bottle was delivered on September 19, 1977, and reported in the Federal Register. Extracts appeared in *CHEMTECH*, November, 1977, p. 671. Monsanto's dissatisfaction with the Delaney Amendment is expressed in a pamphlet, *The Chemical Facts of Life*, available from Monsanto Co., 800 N. Lindbergh, St. Louis, MO 63166.

Sec. 6.4.7   There is a whole literature on hazard analysis. The student might like to start with the papers in session 5 of the *Symposium on Chemical Process Hazards*, **5** (1974), 280, 1. Chem. E. Rugby, England, and with T. A. Kletz, *I. Chem. E. Symposium*, **34** (1971), 75; *New Scientist* May 12, 1977, p. 320, and *Chem. Eng. Prog.*, September 1978, p. 47.

The recent 15-month extension of the 18-month moratorium on a ban on Saccharin was discussed in *Chem. Marketing Reporter*, May 14, 1979.

Sec. 6.5   Many of the arguments and statistics in this section are drawn from papers presented to the Continental European Section of the Society of Chemical Industry, Vienna, October, 1977, and later published under the heading *Growth of International Chemicals*. The papers were, R. Malpas, "Challenge and Opportunity," *Chem. and Ind.*, (1977), p. 886, from which figures and text are adapted; J. Pinder, "The Future of International Trade and Investment," *ibid.*, p. 898; and R. Landau and A. I. Mendolia, "International Chemical Investment Patterns Reviewed," *ibid.*, p. 902. Taken together they present a more coherent and less simplified view of the economic future of the chemical industry than has been possible here, and a nice collection of contrasting points of view is represented. Landau and Mendolia also provide a useful bibliography.

Sec. 6.5.2   The Norsk Hydro venture at Rafnes is reviewed in *Chem. Age*, June.

Sec. 6.5.7   The cost of treating effluents in Europe is analyzed in a report, *Treatment of Aqueous Effluents from the Petrochemical Industry*, prepared by the petrochemicals/ecology group of the European Chemical Industry Federation, CEFIC, 1978.

Sec. 6.7   The 1972 quotation comes from the *Chemical Economy*, *op. cit.* The final quotation comes from Micah 4: 4.

INDEX

ABS resin, 83
"Acceptable risk", 266
Acetaldehyde, acrylic acid from, 72
  chlorination of, 61-62
  manufacture of, 57, 60, 61-62, 124
Acetic acid, 228, 254
  in acrylic acid synthesis, 72-73
  manufacture of, 60, 61-62, 124
  from methanol, 62, 124, 231
Acetic anhydride, manufacture of, 61-62
Acetobacter suboxydans, 155, 156
Acetone, bisphenol A from, 95-96
  production of, 70, 74-76
  from propylene, 74-75
  reaction with iodine, 226
  uses of, 70, 74-76
Acetylene, acrylic acid from, 72
  from carbide, 111, 131
  isoprene from, 89
  preparation of, 111-112
  reaction with formaldehyde, 112-113
  review of, 126
  sources, 108, 245
  uses, 112, 113
  vinyl chloride from, 58
Acid-base catalysis, 226-228
Acids, as catalysts, 225
  as initiators, 186, 189-190
Acrolein, 70, 71
  conversion to glycerol, 79
Acrylic acid, alternative routes, 72
  production of, 70, 71-73
  uses of, 70, 71-73
Acrylic polymers, 184

Acrylonitrile, adiponitrile from, 74, 85, 125
  by ammoxidation, 73-74, 234
  carcinogenicity of, 264-265
  hydrodimerization of, 74, 125
  hydrolysis to acrylic acid, 71-72
  synthesis of, 70, 73-74
  uses of, 70, 73-74
Acrylonitrile copolymer, for bottles, 264-265
Activation energy, 221, 222
Addition polymerization, 166, 167
Adhesives, properties of, 218
Adipic acid, from butadiene, 91
  preparation of, 97-98
  from tetrahydrofuran, 98
Adiponitrile, from acrylonitrile, 74
  from butadiene, 85-86, 125
  synthesis, 85-87
  from tetrahydrofuran, 98
Aerosols, relation to ozone layer, 262
Alcohols, fatty, 133, 141. *See also* specific
  alcohols, *e.g.* Ethanol, Isopropyl
  alcohol, etc.
Aldehydes, 60-62, 105-106
  by aldol condensation, 61, 80
  by oxo reaction, 224
  see also individual aldehydes
Aldol condensation, 61, 80
Alkali act of 1863, 256
Alkylation, 55, 63, 74, 225
Allophonate, 204
Alumina, 47, 222
Aluminum alkyls, 68
Aluminum triethyl, 68
American Chemical Society, vii

SPECIAL UNITS IN THE CHEMICAL INDUSTRY

PETROLEUM AND REFINERY PRODUCTS

Crude oil and some refinery products are traded in barrels (bbl) of 42 US gallons (gal) ($=35$ Imperial gal). As it is a unit of volume, the weight of a barrel depends on the density of the product. Approximate conversion factors follow:

1000 lb $=$ 3.32 bbl crude oil, 3.83 bbl gasoline, 3.54 bbl kerosene,
    3.40 bbl gas oil, and 3.04 bbl fuel oil.
1 tonne $=$ 7.33 bbl crude oil, 8.45 bbl gasoline, 7.80 bbl kerosene,
    7.50 bbl gas oil, and 6.70 bbl fuel oil.

GASES

Natural gas is measured in standard cubic feet (scf) at 1 atmosphere (atm) and 60°F or in cubic meters (m^3) at 1 atm and 0°C. $1 m^3 = 37.33$ scf; $1 scf = 0.0268 m^3$. Thermal units (heat liberated when a volume of gas is burned) are sometimes used. Calorific values depend on the composition of the gas but are usually 900–1000 Btu scf^{-1}. Accordingly 1 therm $= 10^5$ Btu $= 100$–110 scf.

Other gases are also measured in scf and m^3. If the molecular weight of a gas is M, then 10^6 scf of the gas weigh $2.635M$ thousand pounds. For example, 10^6 scf of hydrogen weigh 5.312 thousand pounds and of oxygen 84.32 thousand pounds. Similarly, 1000 m^3 of a gas weigh $0.0446 M$ tonnes. 1000 m^3 of hydrogen weigh 0.0900 tonnes and of oxygen 1.427 tonnes.